# Advances in Intelligent Systems and Computing

## Volume 306

**Series editor**

Janusz Kacprzyk, Polish Academy of Sciences, Warsaw, Poland
e-mail: kacprzyk@ibspan.waw.pl

*About this Series*

The series "Advances in Intelligent Systems and Computing" contains publications on theory, applications, and design methods of Intelligent Systems and Intelligent Computing. Virtually all disciplines such as engineering, natural sciences, computer and information science, ICT, economics, business, e-commerce, environment, healthcare, life science are covered. The list of topics spans all the areas of modern intelligent systems and computing.

The publications within "Advances in Intelligent Systems and Computing" are primarily textbooks and proceedings of important conferences, symposia and congresses. They cover significant recent developments in the field, both of a foundational and applicable character. An important characteristic feature of the series is the short publication time and world-wide distribution. This permits a rapid and broad dissemination of research results.

*Advisory Board*

Chairman

Nikhil R. Pal, Indian Statistical Institute, Kolkata, India
e-mail: nikhil@isical.ac.in

Members

Rafael Bello, Universidad Central "Marta Abreu" de Las Villas, Santa Clara, Cuba
e-mail: rbellop@uclv.edu.cu

Emilio S. Corchado, University of Salamanca, Salamanca, Spain
e-mail: escorchado@usal.es

Hani Hagras, University of Essex, Colchester, UK
e-mail: hani@essex.ac.uk

László T. Kóczy, Széchenyi István University, Győr, Hungary
e-mail: koczy@sze.hu

Vladik Kreinovich, University of Texas at El Paso, El Paso, USA
e-mail: vladik@utep.edu

Chin-Teng Lin, National Chiao Tung University, Hsinchu, Taiwan
e-mail: ctlin@mail.nctu.edu.tw

Jie Lu, University of Technology, Sydney, Australia
e-mail: Jie.Lu@uts.edu.au

Patricia Melin, Tijuana Institute of Technology, Tijuana, Mexico
e-mail: epmelin@hafsamx.org

Nadia Nedjah, State University of Rio de Janeiro, Rio de Janeiro, Brazil
e-mail: nadia@eng.uerj.br

Ngoc Thanh Nguyen, Wroclaw University of Technology, Wroclaw, Poland
e-mail: Ngoc-Thanh.Nguyen@pwr.edu.pl

Jun Wang, The Chinese University of Hong Kong, Shatin, Hong Kong
e-mail: jwang@mae.cuhk.edu.hk

More information about this series at http://www.springer.com/series/11156

Anis Laouiti · Amir Qayyum
Mohamad Naufal Mohamad Saad
Editors

# Vehicular Ad-hoc Networks for Smart Cities

First International Workshop, 2014

 Springer

*Editors*
Anis Laouiti
Telecom SudParis
Institut Mines-Telecom
Paris
France

Amir Qayyum
Mohammad Ali Jinnah University
Islamabad
Pakistan

Mohamad Naufal Mohamad Saad
Universiti Teknologi Petronas
Perak
Malaysia

ISSN 2194-5357
ISBN 978-981-287-157-2
DOI 10.1007/978-981-287-158-9

ISSN 2194-5365   (electronic)
ISBN 978-981-287-158-9   (eBook)

Library of Congress Control Number: 2014948762

Springer Singapore Heidelberg New York Dordrecht London

Printed on acid-free paper

Springer is part of Springer Science+Business Media (www.springer.com)

# Foreword

It is a great honor for us to welcome you to Kuala Lumpur to participate in the First International Workshop on Vehicular Ad-hoc Networks for Smart Cities. Vehicular communication is a key technology in intelligent transportation systems. For many years, the academic and industrial research communities have been investigating these communications in order to improve efficiency and safety of future transportation. Vehicular networking will offer a wide variety of applications, including safety and infotainment. It is envisioned that future communicative vehicles will evolve in a more intelligent environment, also known as smart cities. In this context, the interaction between vehicles and intelligent infrastructures will influence each other, to achieve their targets. Not only would car drivers travel in an efficient and safe manner but the smart cities would also offer the best living conditions for citizens by reducing air and noise pollutions for the inhabitants, and reduce traffic congestion with a better traffic information system for cars. Efficient infrastructures for interaction between vehicles and smart cities is naturally needed to reach these goals.

IWVSC 2014 aims at providing a forum to bring together people from both academia and industry, to discuss recent developments in vehicular networking technologies and their interaction with future smart cities, in order to promote further research activities and challenges. We hope you will find the technical program and the keynote talk very beneficial.

Producing a conference is always a team effort involving many volunteers and we would like to thank the team that made IWVSC 2014 possible. In particular, we are greatly indebted to our Technical Program Committee members who worked hard to produce a comprehensive, high-quality program. The workshop received 12 submissions of which 7 full papers were accepted for final presentation. In addition, IWVSC features one keynote speech and one invited talk on some important issues for VANETs, namely IPv6 use in VANETs and security issues.

   Last but not least, we are also grateful to all authors for their submissions. We
hope that you will find this program interesting and that the workshop will provide
you valuable opportunities to share ideas with other researchers and practitioners
around the world.

Kuala Lumpur, June 2014                          Mohamad Naufal Mohamad Saad
                                                                Anis Laouiti
                                                               Amir Qayyum

# IWVSC 2014 Workshop Organization

**General Chairs**
    Mohamad Naufal Mohamad Saad, Universiti Teknologi Petronas, Malaysia
    Anis Laouiti, Telecom SudParis, France
    Amir Qayyum, Mohammad Ali Jinnah University, Pakistan

**Program Committee**
    Saadi Boudjit, University of Paris 13, France
    Hakima Chaouchi, Telecom SudParis, France
    Yacine Ghamri, University La Rochelle, France
    Halabi Hasbullah, Universiti Teknologi Petronas, Malaysia
    Anis Laouiti, Telecom SudParis, France
    Saoucene Mahfoudh, Jeddah, Saudi Arabia
    Paul Muhlethaler, INRIA, France
    Amir Qayyum, Mohamad Ali Jinnah University, Pakistan
    Naufal Saad, Universiti Teknologi Petronas, Malaysia
    Ahmed Soua, NIST, USA
    Hajime Tazaki, University of Tokyo, Japan
    Apinun Tunpan, Aintec, Thailand
    Wei Wei, Xi'an University, China
    Rachid Zagrouba, ENSI, Tunisia

**Additional Reviewers**
    M. Ali Aydin, University of Istanbul, Turkey
    Jacques Bou Abdo, University of Notre Dame, Lebanon

# Sponsoring

- Telecom SudParis, Institut Mines-Telecom, France
- Universiti Teknologi Petronas, Malaysia

- Mohammad Ali Jinnah University, Pakistan
- ICT-ASIA (STIC-ASIE) French Regional Program
- ICIAS2014/ESTCON2014 (*International Conference on Intelligent and Advanced System 2014/The World Engineering, Science and Technology Congress 2014*)

# Contents

**Part III    Data Dissemination in Vanet Track**

# Part I
# Vanet MAC Layer Protocols Track

# MAC Layer Challenges and Proposed Protocols for Vehicular Ad-hoc Networks

Saira Andleeb Gillani, Peer Azmat Shah, Amir Qayyum
and Halabi B. Hasbullah

**Abstract** Vehicular ad-hoc network (VANET) is a special form of mobile ad-hoc network (MANET) in which vehicles communicate with each other by creating an ad-hoc network. Before the deployment of VANET, it is necessary to address some important issues of VANET, specially concerning about architecture, routing, mobility, and security. Medium access control (MAC) protocols specify the way in which nodes share the underlying channel. As no standard exists for VANET, the research community has previously used IEEE 802.11a and 802.11b as the MAC layer access technologies. In this paper, we have discussed that the main challenge for VANET safety applications is to design an efficient MAC, so that all safety-related messages can be sent on time and such a protocol should be reliable because human lives are involved in the case of VANET. First, the challenges and requirements of a medium access protocol for VANET are discussed and then a survey of MAC solutions available in the literature to deal with these challenges in VANET is presented.

**Keywords** VANETs · MAC protocols · Safety applications

S.A. Gillani (✉) · A. Qayyum
Center of Research in Networks and Telecom (CoReNeT),
M. A. Jinnah University, Islamabad, Pakistan
e-mail: sairagilani@yahoo.com; syedasairagilani@gmail.com

A. Qayyum
e-mail: aqayyum@ieee.org

P.A. Shah · H.B. Hasbullah
Department of Computer and Information Sciences (CIS),
Universiti Teknologi PETRONAS, 31750 Tronoh, Perak, Malaysia
e-mail: pshah_g01944@utp.edu.my

H.B. Hasbullah
e-mail: halabi@petronas.com.my

© Springer Science+Business Media Singapore 2015
A. Laouiti et al. (eds.), *Vehicular Ad-hoc Networks for Smart Cities*, Advances
in Intelligent Systems and Computing 306, DOI 10.1007/978-981-287-158-9_1

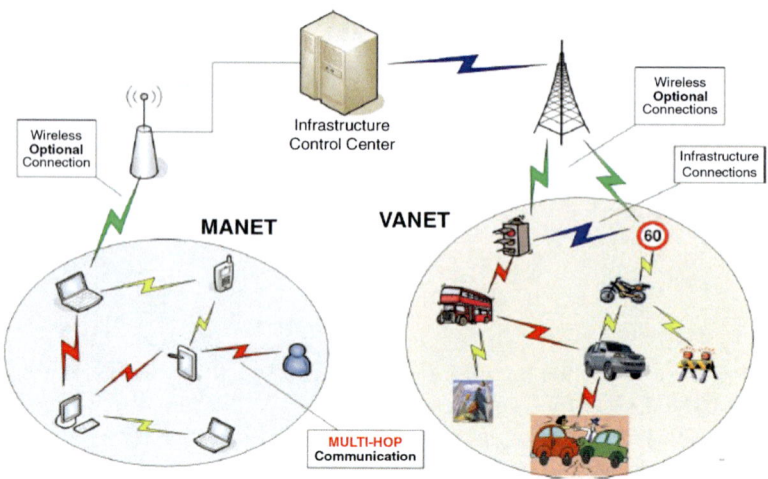

**Fig. 1** MANET versus VANET

# 1 Introduction

Vehicular ad-hoc network (VANET) is a special type of ad-hoc network that is like mobile ad-hoc network (MANET). It can be utilized to improve vehicle safety, enhance traffic efficiency, and provide infotainment in vehicles. VANET has several distinguishing characteristics that differentiate it from MANET. In VANET, topology is very dynamic because vehicles move at a high speed. Blum and Eskandarian [1] stated that multihop paths in VANET are very short-lived because vehicles move at very high speed, as compared to the MANET. Unlike MANET, the mobility of vehicles is regular and predictable in VANET, and there are no power constraints. Vehicles can be equipped with some positioning systems (GPS and GALILEO), through which vehicles' position can be predicted. This predictability allows an improvement in link selection. Figure 1 shows the difference between the MANET and VANET.

VANETs are considered to be a class of MANET. However, VANETs have some distinguishing characteristics. Some of these characteristics, as discussed by Gillani et al. [2], are high node mobility, high dynamic network topology, enough battery power, sufficient storage capacity, high processing power, and availability of GPS. Due to these characteristics of VANET, the medium access control (MAC) solutions presented for MANET are not suitable for VANET. A MAC solution is required to specify the way the nodes share the underlying channel.

While designing a MAC protocol for VANET, the type of messages should also be considered [3]. As there are three types of priority-based messages in VANET applications, hence the MAC protocol should allocate the channel on the basis of message type. The first type of message is periodic messages which give information about the vehicles' current status (position, speed, and direction), these messages are usually needed to broadcast. The second type of messages is of an event-driven messages (emergency message usually related to safety) and these messages have high priority. This type of message is very time-critical, so need high transmission rate. The third type of message is informational messages (non-safety application messages). These messages need prioritized access. In addition to handling the channel access on the basis of message types, a VANET MAC protocol should also consider some other challenges of VANET like its decentralized communication mode (most VANET applications don't rely on any infrastructure) and unpredictable response and reliability.

This paper presents the MAC layer challenges of VANETs and state-of-the art solutions proposed in literature to meet those challenges. The MAC protocols are classified into three classes depending upon the mechanism they use for channel access.

## 2 Applications of VANET

VANETs offer possibilities for new applications, which will make our transportation system secure and efficient. But for various requirements, VANET's applications can be divided into different types. Here, some representative existing applications and several potential future applications of VANET are discussed. Xu et al. [4] and Qian et al. [5] described the categories of VANET's applications as: life critical, safety and warning, e-Toll collection, Internet access, group communications, and roadside service finder. In this work, we have divided VANET's applications into three main categories: safety applications, traffic management applications, and user applications.

### 2.1 Safety Applications

Most desirable group of applications for VANETs is safety applications. To avoid the accidents, road safety applications can play an important role. Even in the case if the accident is unavoidable, these applications can at least minimize the impact of accidents. Safety applications are delay sensitive and they mainly rely on reliable inter vehicle communication. All safety applications require the exchange of messages with other vehicles. The data in these applications are obtained from sensors or other vehicles. The data are processed in each application and after processing it sends messages to nearby vehicles or to the infrastructure. There are two reasons of sending these messages:

1. Awareness of the environment (Periodic Messages)
2. Detection of an unsafe situation (Event-Driven Messages)

An example of safety application is an early warning system. In such type of systems, a driver can be alerted about the road situation, for example, there is an accident on the road ahead, thus giving enough time to the driver so that he can apply the brakes well in time before hitting the accidental car ahead. According to Wang and Thompson [6], if half a second before the collision the warning message is given to the driver, then more than 50 % of the road accidents can be avoided. With the help of early warning system, this number can also be reduced. In addition to the car accident warning messages, some other warning systems can also be installed to avoid the accidents, for example, work zone warning, speed breaker warning, low bridge warning for trucks, etc.

As mentioned earlier, the safety applications are strictly delay sensitive. A fraction of a second is important in decision-making. Hence, there is a requirement for hard deadline for message delivery, so some special handling is required at the lower layers of the TCP/IP protocol stack. As a concern of the network layer, the safety applications do not require or involve in routing. It is because the neighbours of the source node are usually the target audiences for these messages. Therefore, there is no need to send these messages to the nodes that are more than one hop away from the source node. Hence, putting the whole burden on the MAC layer for delivery of messages with minimum delay.

## 2.2 Traffic Management

The congested road notification (CRN) is a traffic management application for VANETs. Through this application traffic, congestion on the road can be notified. For route and trip planning, we can use this application. Through this application, road congestions can be controlled and the information about the best route can be provided to a driver with efficient road conditions. In this application, some roadside units like intelligent traffic signals or electronic sign boards can also be involved to capture and to disseminate the information. The traffic congestion information about the road ahead can positively help to reduce the congestion and to improve the capacity of roads.

Another interesting application for traffic management is the e-Toll plaza, where vehicles do not need to stop to pay toll fee. By using intelligent traffic signals, congestion at road intersections can be efficiently handled. In reaction to the traffic situations at an intersection, these traffic signals can adjust themselves and can also communicate the status to neighboring intersections. This information can be displayed on the e-sign boards by these neighboring intersections.

Another traffic management application is parking availability notification (PAN). Through this application, available slots in parking lots can be found.

The majority of the traffic management applications use the roadside infrastructure. It is possible that some infrastructure is free and can be used by any user while some of them will need a subscription and will not be available to all users. For example, e-Toll infrastructure will require a subscription to offer its services. For these applications, the infrastructure needs to be managed and updated. For these applications to work, the infrastructure with relevant information needs to be managed and controlled.

## 2.3 User Applications

In addition to the road safety applications, information and entertainment applications can also play a vital role in pushing new technologies in VANET. Hot spots for transferring maps are one example of user applications. The passengers in a vehicle can enjoy the facility of Internet connectivity where other traditional wireless internet connectivity options (Wi-Fi, Wi-MAX etc.) are not available. Peer-to-Peer applications can also find their place in VANETs, e.g., gaming, chatting, file sharing, etc. But in all these applications, a large amount of data are transferred so there is a need of specialized means of delivery.

# 3  MAC Protocols for VANET

A MAC protocol specifies the mechanism in which nodes share the channel. There are many MAC issues in VANET like prioritized access, unpredictable response, and reliability. All these are needed because these are basic requirements of safety applications. To provide a reliable broadcast communication can be difficult in wireless networks due to hidden terminal and exposed node problems. A key challenge of VANET is that there are frequent changes in the network topology because the vehicles travel at a high velocity. So, VANET MAC protocols have to care about the rapid topology changes, and for safety applications, there is a need to reduce the medium access delay. In this work, we have categorized the MAC protocols for VANET in three categories. These are:

- Contention-Based Protocols
- Delay Bounded/Contention Free
- Hybrid MAC Protocols

This classification of MAC protocols is shown in Fig. 2.

## 3.1 Contention-Based Protocols

In contention-based protocols, nodes that want to communicate compete for the channel access and the node which wins can use the shared medium for negotiated time. However, there is no delay boundness so the real-time delivery of safety messages may not be guaranteed through these protocols. These protocols are suitable in scenarios where network traffic is bursty so that the bandwidth can be utilized effectively. These protocols are not appropriate for multimedia and real-time traffic as well. The efficiency of these protocols is affected when there are large numbers of users.

The carrier sense multiple access (CSMA) is a well-known example of contention-based MAC. In CSMA, to avoid the collision the transmitting device first listens to the network before transmission. In these types of protocols, collision may

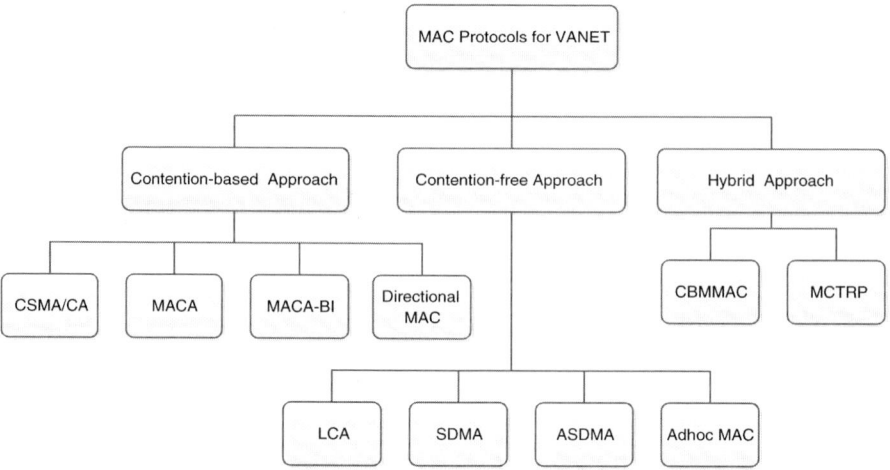

**Fig. 2** MAC protocols for VANET

occur, and due to collisions, packets may suffer unbounded delays. For VANET safety applications, MAC protocols need to reduce the medium access delay.

The IEEE 802.11 standard (802.11a/b/g) is based on CSMA/CA. IEEE 802.11b uses direct sequence spread spectrum (DSSS) as the modulation scheme which reduces multipath fading. IEEE 802.11b/g both are used in many VANET prototypes.

For the purpose of increasing access fairness, Karamad et al. [7] proposed a contention-based MAC for vehicle-to-infrastructure (V2I) communication. This MAC protocol is proposed for the road side unit (RSU) based communication and is based on IEEE 802.11. In this scheme, the distributed coordination function (DCF) is adjusted for node speed. For the fair access of shared medium, an increase in the contention window of each vehicle is done. The proposed approach is not suitable for vehicle-to-vehicle (V2V) communication as it has many limitations that include: requirements for a high coordination, awareness, and overhead that will enable each node to adjust the DCF relative to the other nodes. In addition, the contact time for V2V communication is very short and the speed of other nodes in the communication range is known once only. Hence, the justification of the huge cost of data transfer time for a small benefit of channel access fairness is difficult.

Yang et al. [8] and Pal et al. [9] enhanced the CSMA-based protocols by providing different priority levels. By using different back-off time spacing, packets with higher priorities allow to access the channel faster as compared to the low priority packets. IEEE 802.11e standard also provides different priority levels that allow packets with high priority to access channel fast because the listening period before an attempt is made for channel access is shorter.

Bilstrup et al. [10] discussed that the prioritization mechanism of IEEE 802.11e is also included in IEEE 802.11p, which is forthcoming standard for VANET. This forthcoming standard will use DCF and enhanced distributed channel access (EDCA) as MAC. This standard can play an anchor role for future inter-vehicular communication. But it does not support reliability because in CSMA, nodes compete to access the channel and if the channel is busy, the node must defer its access and waits to attempt channel access for next time and this wait can lead to unbounded delay, which increase the unreliability.

MACA and MACA-BI are also contention-based protocols, so these are also not delay-bounded. Young et al. [11] stated in their work that such protocols (MACA and MACA-BI) are suitable for burst data but they are not suitable for real-time communication. However, when the number of vehicles is increased then the efficiency of these protocols becomes poor.

## 3.2 Contention-Free Protocols

In the contention free or controlled access protocols, like time division multiplexing access (TDMA) and frequency division multiplexing access (FDMA), the access to the medium is pre-allocated. A MAC protocol is contention-free if the nodes do not need to compete for the channel access. Different TDMA-based protocols, as described by Blum and Eskandarian [12], use different methods to assign time slots. Explicit time slot allocation approaches suffer from underutilization of bandwidth. Through this explicit time allocation, MAC may provide guaranteed quality of service (QoS). The drawback of these protocols is that they need a central entity responsible for fair distribution of channel resources among nodes.

Yang et al. [13] proposed a cooperative TDMA MAC protocol for broadcasting. In his paper, the author first described a way to make clusters in VANET. These clusters are based on vehicles directions, i.e., all vehicles are moving in one direction will be a part of one cluster. In each cluster, a node is selected as a cluster head (CH) and other nodes are cluster members (CM) and all they can communicate directly to CH. When any CM notices an accident, it informs to CH. The CH consolidates all messages that come from CM and then broadcast to all its members. The reception of message is verified by ACK. Every receiving member adds ACK in the next frame. If any member does not receive this message, then any member node that has received that message can be acted as a potential helper and maintains a list of all nodes that have not received that message and send this list to CH. The CH finally selects the helper node that will perform cooperation in idle slots for all unsuccessful members. In this way, reliability of safety messages can make possible. The authors did not discuss different priority messages, they just consider safety messages, but to design any MAC protocol for VANET it should be described how channel switching will occur because VANET performs a multichannel operation.

Wang et al. [14] presented a multichannel coordination MAC protocol for contention free access to service channels. This protocol dynamically adjusts the length of intervals between control channel and service channels based on traffic loads of each channel. For this purpose, they used a Markov chain model to find stationary probability and then finally derived the optimum ratio between control channel interval and service channel interval. In this paper, for calculating optimum control channel interval, the authors used the total number of nodes that are sending safety messages, but they did not explain how they will calculate total number of such nodes in dynamically changing traffic conditions in VANET environment. The authors also did not explain how their protocol will avoid interference from hidden nodes.

CDMA is based on spread spectrum (SS) technique that allows several transmitters to send information simultaneously over a single communication channel. Nagaosa and Hasegawa [15] proposed a multi-code sense (MCS) protocol for inter vehicle communication. This protocol is based on CDMA system in which vehicle that wants to communicate senses the currently used codes in the network and through this procedure it determines unused codes. When the number of vehicles increases, then the sequence search for free code can take longer time.

Bana et al. [16] proposed a self-organizing architecture division multiple access (SDMA) for MANETs in which they use space division multiplexing (SDM) in which road is divided into pieces that called spaces and in each space TDMA scheme is mapped. Each vehicle will use different time slots and it will depend on where the vehicle is at this time. If we consider applying this scheme in highway scenario where traffic is sparse, then the overall network utilization will be low because many slots will be unused to low traffic.

For VANET environments where network topology changes rapidly, these types of protocols are not suitable because they do not incorporate the dynamics of the network.

Katragadda et al. [17] proposed location-based channel access (LCA) protocol. In this protocol channel, allocation to nodes is based on their current geographical location. The geographical location of nodes can be obtained by GPS or any other geo localization systems. For this protocol, no central device is needed. The geographical area is divided into cells and each cell is allocated a unique channel. Any node located in a given cell will communicate on the channel associated with that cell. At regular intervals, this protocol assigns channel to vehicles on the basis of current location. Hence, the communication in VANET is delay bounded and real-time. Every user gets a fair chance to communicate. In LCA, we can use TDMA, CDMA, or FDMA as multiple access schemes. In inter vehicle communication, mostly applications need for simultaneous reception of multiple channels, so if LCA is using FDMA and CDMA, the receiver complexity will increase.

Borgonovo et al. [18] proposed a new distributed MAC protocol that is the extension of R-ALOHA protocol that avoid hidden-terminal problem. Through this protocol, reliable single hop broadcasting in ad-hoc networks is possible. It is based on UTRA-TDD slotted physical channel, but can use any other physical standard. In this protocol, every active node can reserve the channel by capturing a frame. A terminal that wants to communicate selects a basic channel (BCH) and if it requires using high speed channel then it can reserve more slots. Still no simulation results of this protocol are reported so cannot say how it will perform under high mobility networks.

Flaminio and Antonio [19] proposed AD-HOC MAC, a new, flexible, and reliable MAC architecture for ad-hoc networks that is based on a completely distributed access technique, RRALOHA. It was developed for the CarTALK 2000 project. The AD-HOC-MAC protocol is used where the communicating nodes can be divided into groups or clusters and all the nodes in a group uses broadcast radio communication. At the physical layer, nodes that are present in different clusters will not communicate with each. The drawback of AD-HOC-MAC is that the vehicles in the same communication range have not to be greater than the number of slots in the frame time.

Ko et al. [20] stated that IEEE 802.11 standard is designed for Omni directional antennas so if we use directional antennas for such MAC protocols then we cannot get additional benefits. Kulkarni [19] states that directional antennas can be beneficial for ad-hoc networks because directional antennas offers higher data rates, better connectivity, and Improved spatial channel utilization due to directional communication. The directional antennas are used to reduce the contention, caused by the transmission of multiple nodes, by dividing the area into regions. A single directional antenna is used at the receiver that is directed at a single region, so the transmissions from the nodes of different regions may not interfere. A new directional MAC (D-MAC) scheme that works similar to IEEE 802.11 except that the ACKs are sent using a directional antenna was also proposed.

## 3.3 Hybrid MAC Protocols

A multichannel token ring protocol (MCTRP) for VANET was proposed by Bi et al. [21]. The objective of this protocol is to develop a MAC method that independently organizes the nodes into token passing rings so that the delay for safety messages can be decreased with an increase in the throughput for other applications. A TDMA-based token passing mechanism is used to control the medium access for intra ring transmissions. While, for inter ring transmission, emergency messages, and for ring administration transmissions CSMA/CA is used to control medium access.

There are many shortcomings in the MCTRP. These include the nodes that founded the ring topology are the main nodes and it makes the system severely dependent on these founder nodes. If any of the leader nodes have problem or becomes unreachable in the topology, then the whole ring collapses. This requires the re-initiation of the ring association. Such type of MAC is more suitable for the scenarios where vehicles move in a platoon form. For the token passing and safety-related messages, CSMA is used by the MCTRP, while data transfer is done through TDMA. In case of heavy loads, this will lead to scenarios where an unbounded delay can be faced by the safety messages. The system also requires GPS for external timing and for location information. This protocol requires two radios per vehicle and for each radio a complete transmission slot is used that may result in inefficient transmission as a node may not need to transmit.

A clustering-based multichannel MAC (CBMMAC) similar to the MCTRP [21] was proposed by Su et al. [22]. A cluster-based technique is used in which the cars in close proximity and traveling in the same direction are grouped into clusters.

A self-elected CH is chosen that controls the cluster. Each node is equipped with two transceivers, one using IEEE 802.11-based contention and the other using contention-free TDMA. The CBMMAC is intended for highway traffic and the system is only activated when a vehicle enters a highway. The approach is dependent upon a CH that relies on vehicles grouping at low levels of relative mobility. An advantage that is diminished by using this technique is the use of all seven channels within one group. In case of one group, high channel utilization will be done. But in the case when two or more groups are in close proximity high collision levels are there. Just like MCTRP, it also requires two radios per vehicle and the GPS.

# 4 Conclusion

In this paper, some basic concepts of VANET were discussed. The applications of VANET were identified and were classified into three classes: safety application, traffic management applications, and user applications. For safety applications, there is a need to reduce the medium access delay as these applications involve human lives. This paper presented a survey of MAC protocols for VANET. The MAC protocols were classified into three categories: contention-based MAC protocols, contention-free MAC protocols, and hybrid MAC protocols.

# References

1. Blum, J.J., Eskandarian, A.: Adaptive space division multiplexing: an improved link layer protocol for inter-vehicle communications. In: IEEE Proceedings of Intelligent Transportation Systems, pp. 455–460. IEEE (2005)
2. Gillani, S., Khan, I., Qureshi, S., Qayyum, A.: Vehicular ad hoc network (VANET): enabling secure and efficient transportation system. Technical Journal, University of Engineering and Technology, Taxila, vol. 13 (2008)
3. Almalag, M.S., Weigle, M.C., Olariu, S.: MAC protocols for VANET. Mobile Ad Hoc Networking: Cutting Edge Directions, 2nd edn, pp. 599–618. Wiley, Hoboken (2013)
4. Xu, Q., Mak, T., Ko, J., Sengupta, R.: Vehicle-to-vehicle safety messaging in DSRC. In: Proceedings of the 1st ACM International Workshop on Vehicular Ad Hoc Networks, pp. 19–28. ACM (2004)
5. Qian, Y., Moayeri, N.: Design of secure and application-oriented VANETs. In: IEEE Vehicular Technology Conference, 2008, VTC Spring 2008, pp. 2794–2799. IEEE (2008)
6. Wang, C.D., Thompson, J.P.: Apparatus and method for motion detection and tracking of objects in a region for collision avoidance utilizing a real-time adaptive probabilistic neural network. U.S. patent no. 5,613,039 (1997)
7. Karamad, E., Ashtiani, F.: A modified 802.11-based MAC scheme to assure fair access for vehicle-to-roadside communications. Comput. Commun. 31(12), 2898–2906 (2008)
8. Yang, S., Refai, H.H., Ma, X.: CSMA based inter-vehicle communication using distributed and polling coordination. In: IEEE Proceedings of Intelligent Transportation Systems, pp. 167–171. IEEE (2005)

9. Pal, A., Dogan, A., Ozguner, F., Ozguner, U.: A MAC layer protocol for real-time inter-vehicle communication. In: Proceedings of the IEEE 5th International Conference on Intelligent Transportation Systems, pp. 353–358. IEEE (2002)
10. Katrin, B., Elisabeth, U., Erik G,S., Urban, B.: On the ability of the 802.11 p MAC method and STDMA to support real-time vehicle-to-vehicle communication. EURASIP J. Wireless Commun. Netw. (2009)
11. Ko, Y.B., Shankarkumar, V., Vaidya, N.F.: Medium access control protocols using directional antennas in ad hoc networks. In: Proceedings of Nineteenth Annual Joint Conference of the IEEE Computer and Communications Societies, INFOCOM 2000, vol. 1, pp. 13–21. IEEE (2000)
12. Blum, J.J., Eskandarian, A.: A reliable link-layer protocol for robust and scalable intervehicle communications. IEEE Trans. Intell. Transp. Syst. **8**(1), 4–13 (2007)
13. Yang, F., Tang, Y., Huang, L.: A novel cooperative MAC for broadcasting in clustering VANETs. In: Proceedings of International Conference on Connected Vehicles and Expo (ICCVE), pp. 893–897. IEEE (2013)
14. Wang, Q., Leng, S., Fu, H., Zhang, Y.: An IEEE 802.11 p-based multichannel MAC scheme with channel coordination for vehicular ad hoc networks. IEEE Trans. Intell. Transp. Syst. **13**(2), 449–458 (2012)
15. Nagaosa, T., Hasegawa, T.: An autonomous distributed inter-vehicle communication network using multicode sense CDMA. In: Proceedings of IEEE 5th International Symposium on Spread Spectrum Techniques and Applications, vol. 3, pp. 738–742. IEEE (1998)
16. Bana, S.V., Varaiya, P.: Space division multiple access (SDMA) for robust ad hoc vehicle communication networks. In: IEEE Proceedings of Intelligent Transportation Systems, pp. 962–967. IEEE (2001)
17. Katragadda, S., Ganesh Murthy, C.N.S., Rao, R., Mohan Kumar, S., Sachin, R.: A decentralized location-based channel access protocol for inter-vehicle communication. In: Proceedings of the 57th IEEE Semiannual Vehicular Technology Conference, 2003. VTC 2003-Spring, vol. 3, pp. 1831–1835. IEEE (2003)
18. Borgonovo, F., Capone, A., Cesana, M., Fratta, L.: RR-ALOHA, a reliable R-ALOHA broadcast channel for ad hoc inter-vehicle communication networks. In: Proceedings of MedHocNet (2002)
19. Borgonovo, F., Capone, A., Cesana, M., Fratta, L.: AD-HOC MAC: new MAC architecture for ad hoc networks providing efficient and reliable point-to-point and broadcast services. Wireless Netw. **10**(4), 359–366 (2004)
20. Ko, Y.B., Choi, J.M., Vaidya, N.H.: MAC protocols using directional antennas in IEEE 802.11 based ad hoc networks. Wireless Commun. Mob. Comput. **8**(6), 783–795 (2008)
21. Bi, Y., Liu, K.H., Cai, L.X., Shen, X., Zhao, H.: A multi-channel token ring protocol for QoS provisioning in inter-vehicle communications. IEEE Trans. Wireless Commun. **8**(11), 5621–5631 (2009)
22. Su, H., Zhang, X.: Clustering-based multichannel MAC protocols for QoS provisionings over vehicular ad hoc networks. IEEE Trans. Veh. Technol. **56**(6), 3309–3323 (2007)

# An Optimal Strategy for Collision-Free Slots Allocations in Vehicular Ad-hoc Networks

Mohamed Hadded, Rachid Zagrouba, Anis Laouiti,
Paul Muhlethaler and Leila Azzouz Saidane

**Abstract** Research in vehicular ad-hoc networks (VANETs) have attracted a lot of attention in the recent years as emerging wireless technologies have opened up the way to many new exciting applications. VANETs are highly dynamic wireless networks that are designed to support vehicular safety, traffic management, and user-oriented applications. Each vehicle can exchange information to inform other vehicles about the current status or a dangerous situation such as an accident. Detecting and sending information about such situations requires a reliable broadcast service between vehicles, thus increasing the need for an efficient medium access control (MAC) protocol. In this paper, we propose ASAS, an Adaptive Slot Assignment Strategy, which takes advantage of bandwidth spatial reuse and reduces intra-cluster and inter-cluster message collisions without having to use an expensive spectrum and complex mechanisms such as CDMA or FDMA. Cluster heads (CHs) which are elected among the vehicles are then responsible for assigning time slots to the other vehicles in their clusters. The evaluation results show the interest of ASAS in terms of slot reuse and collision rates in different speed conditions.

M. Hadded (✉) · R. Zagrouba · L.A. Saidane
RAMSIS Team, CRISTAL Laboratory, 2010 Campus University, Manouba, Tunisia
e-mail: mohamed.haddad@ensi.rnu.tn

R. Zagrouba
e-mail: rachid.zagrouba@cristal.rnu.tn

L.A. Saidane
e-mail: leila.saidane@ensi.rnu.tn

A. Laouiti
TELECOM SudParis, CNRS Samovar, UMR 5157, Evry Cedex 91011, France
e-mail: anis.laouiti@it-sudparis.eu

P. Muhlethaler
INRIA, BP 105, 78153 Le Chesnay Cedex, Paris, Rocquencourt, France
e-mail: paul.muhlethaler@inria.fr

© Springer Science+Business Media Singapore 2015
A. Laouiti et al. (eds.), *Vehicular Ad-hoc Networks for Smart Cities*, Advances in Intelligent Systems and Computing 306, DOI 10.1007/978-981-287-158-9_2

**Keywords** VANET · Quality of service · MAC protocols · CDMA · TDMA · FDMA

# 1 Introduction

VANETs are based on the combination of ad-hoc and cellular technologies to provide a centralized architecture for vehicle to infrastructure communications (V2I, I2 V) and a decentralized architecture for Vehicle to Vehicle communications (V2 V). Due to the importance of V2 V communications, several research projects are underway to standardize V2 V communication in Europe and around the world such as the Car2Car consortium [1] which seeks to improve road safety, FleetNet [2] is a European project aiming to standardize several solutions in order to ensure the safety and comfort of passengers. In the USA, the Federal Communication Commission (FCC) [3] established the Dedicated Short Range Communications service (DSRC) in 2003. The DSRC [4] radio technology is defined in the frequency band of 5.9 GHz a total bandwidth of 75 MHz. This band is divided into seven channels of 10 MHz for each one. These channels comprise one control channel (CCH) and six service channels (SCHs), each one offering a throughput from 6 to 27 Mbps. The CCH is not only reserved for the network management messages, but is also used to transmit messages of high priority messages. The six SCHs are dedicated to data transmission.

Communication uses beacon messages (current status, aggregate data, and emergency messages). If several vehicles simultaneously broadcast messages, then a collision occurs. It is important to avoid collisions on the CCH in order to ensure a fast and reliable delivery of safety messages. To provide a QoS and reduce collisions on the CCH, we introduce an adaptive slot allocation strategy (ASAS) that takes into account the specificity of VANET networks. The strategy proposed operates at the CHs which are the responsible for assigning disjoint sets of time slots to the members of their clusters according to their directions and positions. Thus, by using a centralized means of slot reservation, we ensure an efficient utilization of the time slots and thereby decrease the rate of merging collisions [5] and hidden node collisions caused by vehicles moving in opposite directions.

The rest of this paper is organized as follows. Section 2 reviews related work on MAC protocols in VANETs. Section 3 sets out the challenges of TDMA based MAC solution deployment. Sections 4 and 5 describe the system and the network model, respectively. We give a detailed description of ASAS in Sect. 6. Conclusion and perspectives are presented in Sect. 7.

# 2 Related Work

Various MAC protocols have been proposed for VANETs based, either on contention-based medium access method CSMA/CA such as IEEE 802.11p [6], or on contention-free medium access schemes using time division multiple access

(TDMA), such as AD-HOC MAC [7], VeMAC [5], or on hybrids of these two methods such as DMMAC [8].

The IEEE 802.11p [6] recently designed by TGp Task Group of IEEE [9] improves the standard IEEE 802.11 to support VANETs. This standard improves QoS by offering different message priorities. The prioritization is achieved by using the Enhanced Distributed Channel Access EDCA [6] technique. However, the IEEE 802.11p standard is a contention-based MAC methods that cannot provide a bound on access delays, which is critical for high priority safety applications [10].

VeMAC [5] is a contention-free medium access control protocol recently proposed for VANETs. The protocol implements multichannel TDMA mechanisms, which reserve disjoint sets of time slots in the CCH for vehicles moving in opposite directions and for road side units (RSU). In VeMAC, each node has two transceivers; the first one is always tuned to the CCH whereas the second one can be tuned to any service channel. The assignment of time slots to vehicles on the CCH is performed in a distributed way in which each vehicle randomly acquires an available time slot, and the assignment of time slots on the SCHs is performed by the service providers in a centralized way. However, the size of each VeMAC packet transmitted by a vehicle on the CCH is large (Vehicle ID, current position, set of one-hop neighbors, and the time slot used by each node within the one-hop neighborhood), which increases the overhead of the VeMAC protocol on the CCH. In addition, its random slot assignment technique is inefficient due to the appearance of free slots.

The proposal in [8] is called the dedicated multi-channel MAC (DMMAC) protocol. The DMMAC architecture is similar to WAVE MAC with the difference that in DMMAC, the CCH Interval is divided into an adaptive broadcast frame (ABF) and a contention-based reservation period (CRP). The ABF period consists of time slots, and each time slot is dynamically reserved by an active vehicle as its basic channel (BCH) for collision-free delivery of the safety message or other control messages. The CRP uses CSMA/CA as its channel access scheme. During the CRP, vehicles negotiate and reserve the network resources on SCHs for non-safety applications. However, the simulation model used to evaluate DMMAC fails to take into account the RSU, velocity variation, joining/leaving of vehicles, and bidirectional traffic. Moreover, it was limited to the case of a straight highway scenario with a number of slots smaller than the maximum number of vehicles in the network, meaning that the number of time slots available is always sufficient for the number of vehicles involved.

## Problems

The first aim of MAC protocols for VANETs is to ensure that each vehicle the time to send messages without collisions. TDMA is a method that can be used to assign one-time slot to each active vehicle. We study below the challenges of MAC solution in VANETs focusing particularly on the TDMA techniques.

## 3.1 Distributed TDMA Slot Allocation

When a distributed scheme is used to allocate a time slot, two types of collision on time slots can occur [11]: access collisions between vehicles trying to allocate the same available time slots, and merging collisions between vehicles using the same time slots.

**Access collision** [5] occurs when two or more vehicles within the same two-hop neighborhood set attempt to access the same available time slot. This problem is likely to happen when a scheme way is used to allow the vehicle to reserve a time slot.

**Merging collision** [11] is a basic problem for mobile ad-hoc networks, this problem occurs when two vehicles in different two-hop sets using the same time slot become members of the same two-hop set due to their mobility. Generally in VANET, merging collisions are likely to occur in the following cases:

- Vehicles moving with different speeds,
- Vehicles moving in opposite directions,
- There is an RSU installed along the road.

## 3.2 Centralized TDMA Slot Allocation

When a centralized scheme is used to allocate a time slot, an inter-cluster interference problem can arise. There are two types of inter-cluster interference [12]: One-Hop neighboring Collision and Hidden Node Collision.

**One-hop neighboring collision** (OH-Collision) occurs when a time slot is used by two neighboring vehicles belonging to neighboring clusters. Figure. 1 shows an example of an OH-collision situation when vehicle CM-31 in cluster III and vehicle CG-45 in cluster IV are using the same time slot. Since CM-31 and CG-45 are within transmission range of each other, then a collision will occur at vehicle CM-31 and CG-45.

**Hidden node collision** (HN-Collision) occurs when two vehicles are in range to communicate with another node, but not within transmission range of each other. Let us consider a situation in Fig. 1 when vehicle CM-31 in cluster III and vehicle CM-44 in cluster IV are using the same time slot. Since these two vehicles are outside transmission range of each other, a collision will occur at vehicle CG-45 of cluster IV.

## 4 Network Model

A VANET network in a highway environment consists of a set of vehicles moving in opposite directions on two roads, each road having two lanes. The vehicles belong to self-organized groups called "clusters." In each cluster, there are three different vehicle states: CH, cluster member (CM), and cluster gateway (CG).

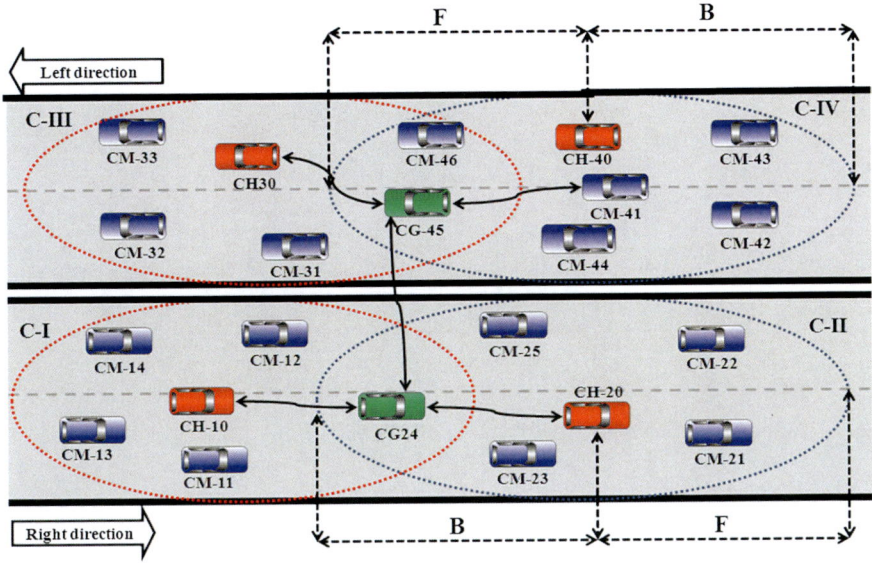

**Fig. 1** Network model

Sometimes, there is another state named undecided state (US) that is used for the initial state of a vehicle. All CMs and CGs are within one hop communication range of the CH, see Fig. 1.

Each cluster has two sets of vehicles: $F$ (Front) and $B$ (Behind).

- $B$ is a set of vehicles that are behind the CH,
- $F$ is a set of vehicles that are ahead of the CH.

Let $C_i$ be a cluster of size $m$ with its cluster head $CH_i$ defined by the position $(x, y, z)$.

$$F_i = \{V_{i,1 \leq i \leq m}(x', y', z'), x' \geq x\}$$

$$B_i = \{V_{i,1 \leq i \leq m}(x', y', z'), x' < x\} = C_i - F_i$$

After the clusters have been set up and the cluster heads have been elected as shown in Fig. 1, each cluster head maintains a local TDMA MAC frame. After a cluster member CM receives its slot allocation from its cluster head, it transmits safety or control messages only during this slot and receives safety messages during other time slots.

# 5 System Model

A vehicle is said to be moving in a left (right) direction if it is currently heading in any direction from North/South to West (East), as shown in Fig. 1. Based on this definition, if two vehicles are moving in opposite directions on a two-way road, it

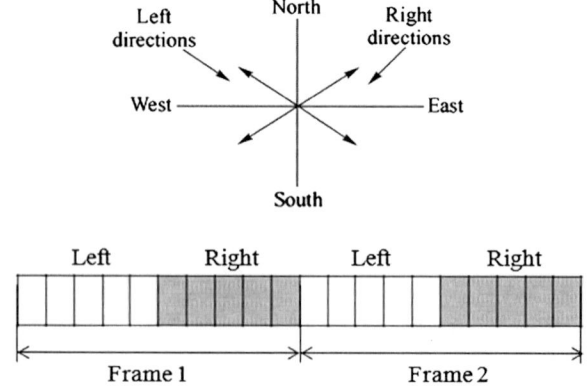

**Fig. 2** Partitioning each frame into two sets: *left* and *right*

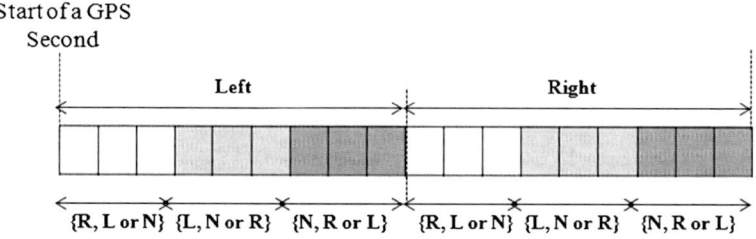

**Fig. 3** Partitioning of each set into $R$, $L$ and $N$

is certain that one vehicle is moving in a left direction while the other vehicle is moving in a right direction [5]. The access time is partitioned into frames and each frame is partitioned into two sets of time slots: *Left* and *Right*, as shown in Fig. 2. The *Left* set is associated with vehicles moving in left directions, while the *Right* set is associated with vehicles moving in right directions.

In ASAS, we assume that each set of time slots *Right* or *Left* is partitioned into three subsets of time slots: $L$, $R$ and $N$, as shown in Fig. 3.

- $L$ is the subset of time slots reserved for vehicles belonging to the $F$ set of vehicles,
- $R$ is the subset of time slots reserved for vehicles belonging to the $B$ set of vehicles,
- $N$ is the subset of unused time slots, in which all vehicles in cluster remain inactive.

## 6 ASAS Description

The ASAS protocol is based on a TDMA method, in which the medium is divided into frames and each frame is divided into time slots. Only one vehicle is allowed to transmit in each time slot. This strategy is centralized in stable cluster heads that continuously adapt to a highly dynamic topology. The main idea is to take the direction and position of the vehicles into consideration in order to decide which slot should be occupied by which vehicle. The allocation of time slots is based on requests from the vehicles in their HELLO messages, which are used by the cluster head to calculate the transmission schedule. The strategy is robust in the sense that it provides an efficient time slot reservation without intra-cluster and inter-cluster interferences. In this section, we address two important challenges: cluster formation and the TDMA slot assignment mechanism for intra-cluster and inter-cluster communications.

## 6.1 Cluster Formation

Clustering is the process that divides all the vehicles in a network into organized groups called clusters. Several algorithms such as [13] and [14] have been proposed for cluster formation that take into account the specific characteristics of VANETs. We propose a cluster formation algorithm based on information of the vehicles' position and direction, and which uses the Euclidean distance to divide the vehicles into clusters. To provide more stable clusters, our cluster formation scheme takes into account the direction of the vehicles, i.e., only vehicles moving in the same direction can be members of the same cluster. If the direction is not taken into account in a highway environment with two ways, the vehicles that are moving in opposite direction to the cluster head will only be part of the cluster for a very short time, and a new cluster will have to be formed almost immediately. Through the Euclidean distance and transmission range (i.e., the DSRC range is 1 km), we can decide whether two vehicles can be grouped into the same cluster.

**Cluster head election** Initially, all vehicles are in the USA. To divide the network into clusters, each vehicle broadcasts its current state "position, speed" to notify its presence to its one-hop neighbors. Then, based on the received messages each vehicle can build its one-hop neighboring list. To determine the most stable CH, the election of a cluster head is based on the following function. The elected cluster head is a vehicle which has the minimum average distance to the other vehicles in the cluster, the closest speed to the average speed and the maximum number of neighboring vehicles.

$$F(i) = \alpha \left( \sum_{j \in N(i)} d(P_i, P_j) \right) / n_i + \beta \left( \sum_{j \in N(i)} |V_i, V_j| \right) / n_i - \sigma \times n_i$$

where

$$\begin{cases} n_i: \text{Number of vehicles within one} - \text{hop range of vehicle } i \\ d(P_i, P_j): \text{Euclidean distance between vehicles } i \text{ and } j \\ |V_i, V_j|: \text{Velocity differences between } i \text{ and } j \\ N_i: \text{The set of one hop neighbors of vehicle } i \\ \alpha, \beta, \sigma: \text{The weight coefficients}, \ \alpha + \beta + \sigma = 1 \end{cases}$$

The vehicle that has the minimum value of $F$ is elected as the CH. All the vehicles that are within transmission range of the CH become CMs and are not allowed to participate in another cluster head election procedure until it becomes necessary.

$$\begin{cases} CH = \{i/F(i) = \text{Min}(F(j), \forall j \in N(i))\} \\ CM = \{j, \forall j \in N(i) \text{ and } j \neq i\} \end{cases}$$

Once the cluster head has been elected, it starts to periodically broadcast invite-to-join ITJ message to its one-hop neighbors. If a CM receives an ITJ message from another neighboring CH moving in the same direction, it will attempt to get the attention of the CH by sending to it a request-to-join (RTJ) message. Upon the receipt of an ACK message from CH, the corresponding vehicle will switches from CM state to cluster gateway state CG.

**Cluster Maintenance** In VANETs, a vehicle can join or leave a cluster at any time. These two operations will have only local effects on the topology of the cluster if the vehicle is a CM. However, if the vehicle is the CH before leaving the cluster, it must hand over the responsibility to one of the very close CMs. The first reason for that is to keep the cluster organized as a "one-hop cluster with two sets of vehicles F and B" even if the current CH leaves. The second reason is to avoid using the re-clustering algorithm and thus no re-clustering overhead is generated when the cluster head leaves the cluster. Then, the current CH will order the CM to switch to CH and switch its own state to CM.

When the CH receives an ITJ message from another neighboring cluster head moving in the same direction, only one of them will keep its CH responsibility while the other will switch to CM. The CG between two clusters becomes a CM of the new cluster and each CM of the cluster whose CH will become a CM will switch to CM if it receives an ITJ message from the new CH and will switch to US otherwise. Selecting the cluster head when two clusters merge is done according to the function defined in Sect. 6.1.

## 6.2 TDMA Slot Assignment Mechanism

In this study, we assume that each vehicle is equipped with a positioning system, e.g., global positioning system (GPS), which can provide an accurate real-time three-dimensional position (latitude, longitude and altitude), direction, velocity, and exact time. The synchronization between vehicles can be performed by using GPS timing information.

We also assume that the TDMA frame consists of $k$ time slots and each frame is divided into two sets of time slots of size $k/2$, see Fig. 3. The first and the second sets are used by vehicles moving in right and left directions, respectively. However, if a vehicle moving in a right/left direction detects that no free time slot is available for vehicles moving in that direction, then it will request a time slot normally reserved for vehicles moving in the opposite direction. This technique is used to mitigate the merging collision problem. We also assume that each cluster

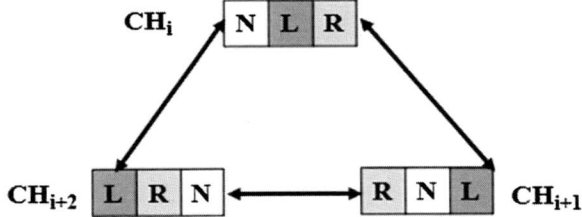

**Fig. 4** Distribution of three time slots subsets

| CH-ID | MAP | | | Size | Slot Status SS_[1] | ...... | Slot Status SS_[K/2] |
|---|---|---|---|---|---|---|---|
| | R, L or N | L, N or R | N, R or L | | Free, VehID or N | | Free, VehID or N |

**Fig. 5** Frame information

head CH maintains two sets of vehicles, F and B, and each of which is divided into three subsets of time slots $L$, $R$, and $N$.

To avoid inter-cluster interference, the allocations of the time slot subsets are different between neighboring clusters (as shown in Fig. 4).

**TDMA Slot Reservation** In this section, we provide a detailed description of our TDMA slot allocation strategy. When a vehicle $V$ needs to access the network, it first sends a reservation request to the cluster head CH for a periodic time slot. When the CH receives the reservation request and depending on the vehicle position, it determines whether the current time slot belongs to the $L$ or $R$ set and then it selects to $V$ the first available slot as its owner slot. Each cluster head CH determines its distribution of three subsets of time slots "MAP" according to the MAPs of their neighboring clusters. The CH can obtain the MAP information of the neighboring cluster heads through the cluster gateways CGs. Once a CH has selected a time slot for a CM, it sends a reservation which includes the slot identifier. However, ASAS requires that every CH should periodically send frame information FI to its two neighboring cluster heads via its CGs (see Fig. 5). This FI contains the following fields:

1. CH-ID, indicates the identifier of CH that sends the FI packet.
2. MAP {{$R$, $L$, $N$}, {$L$, $N$, $R$} or {$N$, $R$, $L$}}.
3. The sizes of $R$, $L$, $N$ subsets.
4. The state of each time slot reserved for the cluster head's moving direction.

The second information element is transmitted only once time and the third is transmitted if the cluster head updates the size of the $L$, $R$, or $N$ subsets. Unlike other slot reservation techniques based on FI broadcasts where each vehicle must determine the set of time slots used by all the vehicles within its two-hop neighborhood in order to acquire a time slot.

| | 1 | 2 | 3 | 4 | 5 | 6 | 7 | 8 | 9 | 10 | 11 | 12 | 13 | 14 | 15 | 16 | 17 | 18 | 19 | 20 | 21 | 22 | 23 | 24 |
|---|---|---|---|---|---|---|---|---|---|---|---|---|---|---|---|---|---|---|---|---|---|---|---|---|
| 1 | 10 | 13 | 14 | | 11 | 12 | | | N | N | N | N | | | | | | | | | | | | |
| 2 | 21 | 22 | | | N | N | N | N | 20 | 23 | 24 | 25 | | | | | | | | | | | | |
| 3 | | | | | | | | | | | | | 30 | 32 | 33 | | 31 | | | | N | N | N | N |
| 4 | | | | | | | | | | | | | 41 | 42 | 43 | | N | N | N | N | 40 | 44 | 45 | 46 |

Fig. 6 An example of slot assignment

In our reservation technique, the CH discovers the available slots while requiring less overhead than the other techniques. Moreover, the CH also knows all the time slots which are likely to cause a collision at the transmission channel (i.e., $N$ set). As shown in Fig. 6, especially in the frame information of cluster head number 2 (FI-2), there are two available time slots for new vehicles moving ahead of the cluster head and the reservation of any time slot whose identifier belongs to [5...8] may cause a collision. When all the slots in the $L$ or $R$ subsets are busy, the CH must communicate with its two neighboring cluster heads to reserve a time slot in the $N$ set for new vehicles respectively belonging to the $F$ or $B$ set.

**Property 1** For each cluster $C_i$:

- $\forall k \in R_i$, if $\exists x \in B_i$ such that $FI_i\_SS[k] = x$, thus it is certain that vehicle $x$ can transmit during slot $k$ without intra-cluster and inter-cluster interference from another vehicle.
- $\forall k \in L_i$, if $\exists x \in F_i$ such that $FI_i\_SS[k] = x$, thus insuring that the vehicle $x$ can transmit during slot $k$ without intra-cluster and inter-cluster interference from another vehicle.

**Property 2** For each cluster $C_i$:

- $\forall x \in B_i, \exists k \in R_i$ such that $FI_i\_SS[k] = x \Leftrightarrow |B_i| \leq FI_i\_Size[R]$.
- $\forall x \in F_i, \exists k \in L_i$ such that $FI_i\_SS[k] = x \Leftrightarrow |F_i| \leq FI_i\_Size[L]$.

where $|B_i|$ and $|F_i|$ are respectively the number of vehicles in the $B_i$ and $F_i$ set.

Otherwise, if $|B_i| > FI_i\_Size[R](|F_i| > FI_i\_Size[L])$, i.e., there are vehicles that cannot acquire a time slot, because all the slots in $R_i$ or $L_i$ are busy, in this case the $CH_i$ will communicate with neighboring cluster heads to allocate time slots to these vehicles by shortening the length of $N_i$ and increasing the length of $R_i$ or $L_i$. Then, the cluster head $CH_i$, must update its frame information $FI_i$ and transmit this frame to its two neighboring cluster heads $CH_{i-1}$ and $CH_{i+1}$.

Time slots are allocated according to the vehicle's movement and position. By using a centralized approach, we change the slot allocation process from random reservations to optimal allocations, which can improve the convergence performance of the MAC protocol and achieves provides an efficient broadcast service for the successful delivery of real-time safety information.

**Release of TDMA slots** If, after a specific time, a cluster head does not receive a beacon message from CM to signal its presence, then the CH immediately releases the time slot allocated to the CM and it removes this CM from it cluster (i.e., the $F$ or $B$ set).

**Dynamically reallocating slots** In VANETs especially in a highway environment, the number of vehicles is not equally distributed in each direction. Thus, the ratio between the two slot sets *Left* and *Right* should be adjusted according to the vehicle density. We use the algorithm presented in [15] to adjust the ratio of the two slot sets. In order to describe the ratio adjustment algorithm, the following notations are introduced and valid for a specific moment in time $t$ and for a specific cluster head $n$.

| | |
|---|---|
| $N^n(t)$ | The current CMs of cluster head $n$ |
| $S^n(t)$ | The frame length of the cluster head $n$, i.e., the number of time slots of each frame of cluster head $n$ |
| $S_d^n(t)$ | The number of time slots reserved for the direction $d$ of the cluster head $n$, i.e., *Left* or *Right* set |
| $\rho\text{max}, \rho\text{min}$ | The maximum threshold and minimum threshold, which is a ratio between the number of vehicles in the cluster and the number of slots |

Initially we suppose the number of vehicles in each direction to be equal. The density of vehicles changes as the vehicles move and we reach the conditions expressed in (1), see below. We need to adjust $S_d^n(t)$ to come back to the conditions expressed in (1).

$$\begin{cases} \frac{N^n(t)}{S_d^n(t)} > \rho\text{max} \text{ or } \frac{N^n(t)}{S_d^n(t)} < \rho\text{min} \\ \frac{N^{th}(t)}{S_d^n(t)} < \rho\text{max} \text{ and } \frac{N^{th}(t)}{S_d^n(t)} > \rho\text{min} \end{cases} \tag{1}$$

The cluster head $n$ sends its neighboring cluster heads, through the cluster gateways a proposal to redistribute the number of the *Left* or *Right* slots in the FI. If the neighboring clusters head agree to the proposal, the new slot allocation scheme will be adopted by the neighboring cluster heads that have the same frame length in the next time frame. Each cluster head must save information about how the slots are allocated and periodically sends it in the FI.

## 7 Performance Evaluation

In this section, we evaluate the effect of transmission range and speed variation and we carry out a comparison of ASAS with the DMMAC and VeMMAC protocols.

## 7.1 Simulation Setup

We did several tests to evaluate the performance of ASAS under various realistic conditions. We used VanetMobiSim [16] to create a mobility scenario and we used a JAVA simulation, using the JDK compiler.

**Table 1** The average and the standard deviation of the speed

| $\mu$ (km/h) | $\sigma$ (km/h) |
|---|---|
| 80 | 20 |
| 100 | 30 |
| 120 | 35 |

**Table 2** System parameters for simulation

| Parameter | Value |
|---|---|
| Highway length | 2 km |
| Directions | 2 |
| Lanes each way | 2 |
| Lane width | 5 m |
| Transmission range/Scenario | {150, 350, 550, 750, 1,000} m |
| Slots/ABS frame | 50 |
| Slots for right direction | 25 |
| Slots for left direction | 25 |
| Slot duration | 1 ms |
| Simulation time | 120 s |
| Number of vehicles/Scenario | 60 |

## 7.2 Mobility Scenarios and Simulation Parameters

The mobility scenarios implemented for the highway are with two-way traffic and different density levels in each direction, see Fig. 3. The vehicles are moving at different speeds (Table 1) and have different transmission ranges. During simulation time, each vehicle moves at a constant speed, and the number of vehicles on the highway remains constant. Table 2 summarizes the simulation parameters .

## 7.3 Performance Metrics and Simulation Results

We evaluate our MAC protocol using the following performance metrics:

- MR-Collision rate: the MR-Collision rate is defined as the average number of merging collisions.
- AC-Collision rate: the AC-Collision rate is computed as the average number of access collisions.
- IC-Collision rate: The IC-Collision rate is defined as the average number of inter-cluster collisions due to HN-Collision and OH-Collision. However for DMMAC and VeMAC, the IC-collision rate is defined as the rate of collision between the adjacent sets of two-hop neighboring vehicles that is moving in the same direction.

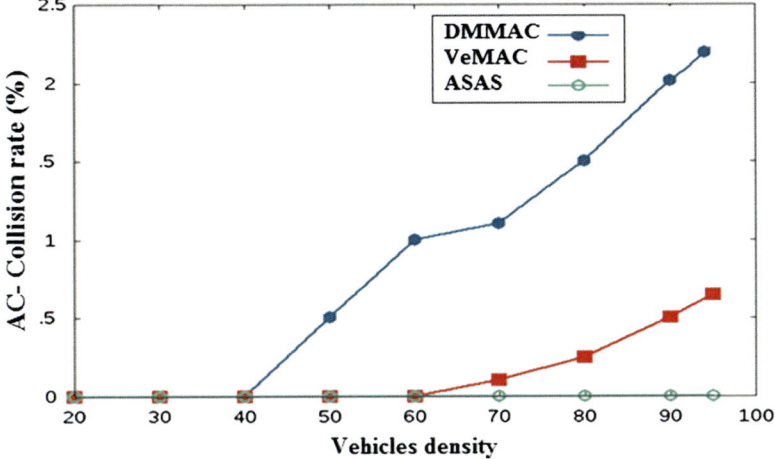

**Fig. 7** The access collision rate as a function of vehicle density

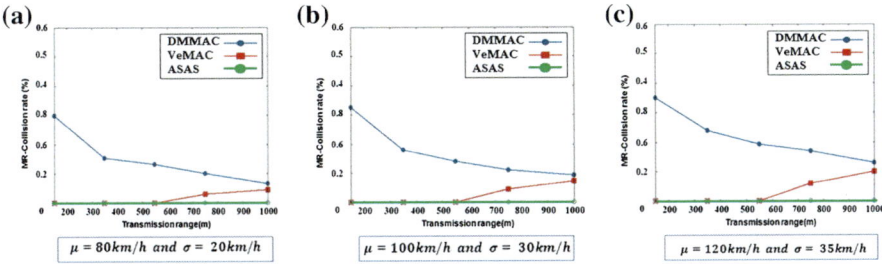

**Fig. 8** The merging collision rate as a function of transmission range

Due to the highly dynamic topology, the number of clusters varies during the simulation time (new cluster are added and clusters are merged) and this variation should be as low as possible. Thus, the cluster formation algorithm proposed reduces the number of new clusters created due to the high mobility of the vehicles. Therefore, it creates stable clusters and keeps the current clusters as stable as possible.

The rate of access collisions under different traffic load conditions is shown in Fig. 7. We note that no access collisions generated by ASAS in contrast to both the DMMAC and VeMAC protocols. The reason is that the assignment of time slots to vehicles is performed by the cluster heads in a centralized manner. The VeMAC protocol generates a higher rate of access collisions than ASAS, especially for a high traffic load but the rate is significantly lower than that generated by the DMMAC protocol. These results show the effectiveness of the ASAS technique.

In Fig. 8, the $x$-axis represents the transmission range, while the y-axis represents the merging collision rate of the vehicles. Figure 8a, b, c shows the rate of merging collisions for DMMAC, VeMAC, and ASAS. It is clear that merging

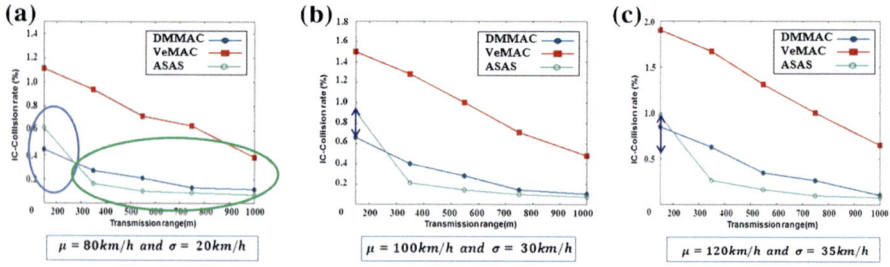

**Fig. 9** The inter-cluster collision rate as a function of transmission range

collisions are entirely eliminated for ASAS as it its merging collision rate is always equal to zero for all velocities and transmission range values. Indeed, ASAS allocates disjoint sets of time slots to vehicles moving in opposite directions. The figure shows also that the merging collision rate is reduced by 100 % compared to the DMMAC and VeMAC protocols. We can see that the ASAS protocol performs even better when the average speed becomes higher and thus the average speed has no impact on the performance of ASAS.

Figure 9 shows the rate of IC-Collisions for the DMMAC, VeMAC, and ASAS protocols. It is clear that ASAS shows a lower rate of IC-Collisions than both DMMAC and VeMAC. The figure shows that the IC-Collision rate is reduced by 50 % compared to VeMAC and by 5–15 % compared to DMMAC. The reason is that ASAS strictly assigns disjoint sets of time slots to vehicles moving behind and ahead of the cluster head. Thus, the protocol decreases collisions between neighboring clusters, which decreases the rate of Inter-cluster collisions compared to the other protocols. We can also see that the IC-Collision rate decreases as the transmission range increases. This is because increasing the transmission range, decreases the number of clusters in the network and thus the inter-cluster collision rate will automatically decrease. We conclude that ASAS can operate successfully under the DSRC architecture because the transmission range in DSRC is equal to 1,000 m. However, we note that if the transmission range is low (less than 250 m) the DMMAC protocol performs slightly better than ASAS. This is due to the large number of clusters which increases the rate of inter-cluster collisions. We can also see that the ASAS performs even better when the average speed is higher.

## 8 Conclusion and Future Work

This paper proposes an ASAS for cluster-based TDMA for VANETs in which the assignment of time slots to vehicles is performed by the cluster heads in order to avoid any access collision problems. ASAS can adapt to different traffic conditions because it has a stable clustering technique that provides stable clusters with less overhead. From the experimental results, we conclude that this strategy achieves an efficient reservation and utilization of the available time slots without access collisions and decreases the rate of merging collisions and inter-cluster collisions caused by the hidden node problem. Compared with the DMMAC and VeMAC protocols, ASAS generates a lower rate of transmission collisions in different transmission ranges, speed scenarios and traffic load conditions. ASAS achieves this without having to use expensive spectrum management mechanisms such as CDMA or FDMA.

In future work, we will study the performance of ASAS in a city scenario and the effect of RSUs on the performance of ASAS. In addition, the dynamic adjustment of the length of the three subsets will be scrutinized. We plan to extend ASAS to support multichannel operation and a reliable broadcast on the control and service channels. We also plan to evaluate ASAS for unicast transmission mode both through simulations and analysis. In addition, we will carry out an experimental comparison with other existing broadcast protocols such as the IEEE 802.11p standard that operates with a DSRC architecture.

# References

1. CAR 2 CAR Communication Consortium. http://www.car-to-car.org/
2. FleetNet homepage. http://www.et2.tu-harburg.de/fleetnet
3. Federal Communications Commission: "FCC 99-305," FCC report and order, Oct 1999
4. The FCC DSRC (Dedicated Short Range Communications) web site. http://wireless.fcc.gov/services/its/dsrc/
5. Zhuang, W., Omar, H.A., Li, L.: VeMAC: A novel multichannel MAC protocol for vehicular ad hoc networks. INFOCOM WKSHPS 413–418 (2011)
6. 802.11p-2010-IEEE standard for information technology—Telecommunications and information exchange between systems—local and metropolitan area networks—specific requirements part 11: Wireless LAN medium access control (MAC) and physical layer (PHY) and physical layer (PHY) specifications amendment 6: Wireless access in vehicular environments (2010)
7. Borgonovo, F., Capone, A., Cesana, M., Fratta, L.: ADHOC MAC: new MAC architecture for ad hoc networks providing efficient and reliable point-to-point and broadcast services. Wireless Netw. **10**(4), 359–366 (2004)
8. Liu, F., Lu, N., Ji, Y., Wang, X.: DMMAC : a dedicated multi-channel MAC protocol design for VANET with adaptive broadcasting. In: Wireless Communications and Networking Conference (WCNC), 1–6, Sydney, Australia (2010)
9. TGp. http://www.ieee802.org/11/Reports/tgp_update.htm
10. Abu-Rgheff, M.A., Abdalla, G.M., Senouci, S.M.: SOFTMAC: space-orthogonal frequency-time medium access control for VANET. In: Information Infrastructure Symposium 20GIIS '09. Global, pp. 1–8, June 2009
11. Borgonovo, F., Campelli, L., Cesana, M., Fratta, L.: Impact of user mobility on the broadcast service efficiency of the ADHOC MAC protocol. Proc. IEEE VTC **4**, 2310–2314 (2005)
12. Wu, T., Biswas, S.: Reducing inter-cluster TDMA interference by adaptive MAC allocation in sensor networks. In: Sixth IEEE International Symposium on a World of Wireless Mobile and Multimedia Networks (WoWMoM'05) (2005)

13. Fan, P., Sistla, P., Nelson, P.C.: Theoretical analysis of a directional stability-based clustering algorithm for vanets. Vehicular Ad Hoc Networks (2008)
14. Shea, C., Hassanabadi, B., Valaee, S.: Mobility-based clustering in VANETs using affinity propagation. In: IEEE Globecom (2009)
15. Wei-dong, Y., Pan, L., Yan, L., Hong-song, Z.: Adaptive TDMA slot assignment protocol for vehicular ad hoc networks. J. China Univ. Posts Telecommun. 11–18 (2013)
16. VanetMobiSim project, home page. http://vanet.eurecom.fr. Accessed 29 May 2010

# Vehicular MAC Protocol Data Unit (V-MPDU): IEEE 802.11p MAC Protocol Extension to Support Bandwidth Hungry Applications

Muhammad Sajjad Akbar, Kishwer Abdul Khaliq and Amir Qayyum

**Abstract** Vehicular ad-hoc networks (VANETs) have been a hot research topic in academia and industry with respect to safety of drivers and entertainment applications. Many MAC layer protocols have defined in research. The IEEE 802.11p is one of the popular carrier sense multiple access (CSMA)-based MAC layer protocol for VANET and is successfully used for safety applications. However, IEEE 802.11p is less efficient for bandwidth hungry and delay-sensitive applications as there is significant fixed overhead of channel access, the inter-frame spaces, and acknowledgments of each frame transmitted. In this paper, an aggregation mechanism, named Vehicular MAC protocol data unit (V-MPDU) with block acknowledgment as an extension of the existing IEEE 802.11p is proposed. The proposed aggregation technique collects frames against each destined node and wraps each frame in a single IEEE 802.11p header. Moreover, it permits each of the aggregated data frames to acknowledge individually or re-transmit in case of any transmission error. Hence, it improves the channel access mechanism in terms of efficiency as multiple frames transmit in single transmission opportunity, which ultimately reduce number of potential collisions and re-transmissions as well. Further, effective bandwidth automatically improves.

**Keywords** Frame aggregation · IEEE 802.11p · VANET · Bandwidth hungry application

M.S. Akbar (✉) · K.A. Khaliq · A. Qayyum
CoReNeT, Mohammad Ali Jinnah University Islamabad, Islamabad, Pakistan
e-mail: sajjad@corenet.org.pk
URL: http://www.corenet.org.pk

K.A. Khaliq
e-mail: kishwer.a.k@ieee.org

A. Qayyum
e-mail: aqayyum@ieee.org

© Springer Science+Business Media Singapore 2015
A. Laouiti et al. (eds.), *Vehicular Ad-hoc Networks for Smart Cities*, Advances in Intelligent Systems and Computing 306, DOI 10.1007/978-981-287-158-9_3

# 1 Introduction

Vehicular ad-hoc network (VANET) is a challenging domain in the wireless networks. VANET can be considered as off shoot of mobile ad-hoc networks (MANETs). IEEE recommends 802.11p as MAC and PHY layer standard to add wireless access in vehicular environments (WAVE) [1, 5, 11], a vehicular communication system which incorporates data exchange between high-speed vehicles (V2V) and between the vehicles and infrastructure (V2I) under licensed ITS band of 5.9 GHz (5.85–5.925 GHz).

In IEEE 802.11p, seven channels of 10 MHz each is available for transmission. There is one control channel (CCH) and six service channels (SCH) in IEEE 802.11p [5, 8, 13]. All the safety applications use the CCH whereas entertainment applications use the SCH. During the transmission of CCH, all the other channels stop their communication and listen to CCH. CCH can use SCH for its communications, but the SCH cannot use the CCH. IEEE 802.11p uses IEEE 802.11e priority mechanisms for QoS support [2, 10] as shown in Fig. 1.

ITS initially gave the concept of safety applications, but the next generation ITS adds the use of bandwidth hungry applications like video on demand (VoD), voice over IP (VoIP), video conferencing, online gaming, file transfer, etc., which require less delay and high bandwidth. Currently, VANET focuses on safety application and uses IEEE 802.11p as MAC and PHY layer standard. At the MAC layer, IEEE 802.11p uses CSMA/CA for channel access. Research shows that IEEE 802.11p performs good for safety application but do not perform well for real time and bandwidth hungry applications in a vehicular environment [2, 4, 6, 12, 16, 17]. Further, under heavy [7] traffic loads there is performance degradation, both for individual nodes and for the whole network due to CSMA. In the worst case, packet drop was near 80 %. So it is unacceptable not only for safety applications, but also for bandwidth hungry applications. The authors have shown through simulations that in 802.11p, backoff window sizes are not adaptive and cause throughput degradation [15]. To handle this issue, in this paper, an aggregation mechanism V-MPDU with block acknowledgment as an extension of the existing IEEE 802.11p is proposed. V-MPDU aggregation collects frames to be transmitted to a single destination and wraps each frame in a single IEEE 802.11p header. V-MPDU permits each of the aggregated data frames to be individually acknowledged or re-transmitted in case of error.

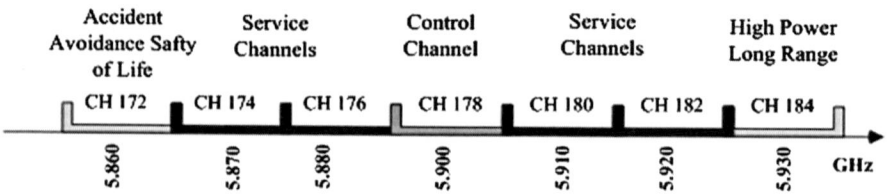

**Fig. 1** IEEE 802.11p channel allocation [2]

## 2 Proposed Extension for IEEE 802.11p

In Internet transmission, most of the packets are small in size that creates overhead specially for wireless networks. Current applications produce large number of small packets in a single burst, and there is significant fixed overhead of channel access, the inter-frame spaces and acknowledgments of each frame transmitted. Traditional MAC schemes do not handle such burst traffic and cause performance degradation in terms of throughput and delay. To reduce this overhead in VANET, a frame aggregation mechanism V-MPDU with block acknowledgment is proposed which transmits two or more frames of the same destination together into a single transmission. It improves the channel access mechanism in terms of efficiency as multiple frames are sent in a single transmission, which ultimately reduces number of potential collisions. Therefore, applications generating burst in VANET environment can be handle efficiently.

### 2.1 Vehicular MAC Protocol Data Units with Aggregation

Here, the structure of the V-MPDU is discussed. We have tried to design the simplest structure so that no extra overhead generates. MAC PDUs are aggregated to produce a single V-MPDU. A delimiter is attached with each MPDU. An aggregated V-MPDU encapsulation is presented in Fig. 2. A delimiter is designed of 24-bits in length, i.e., 16-bits Size field and 8-bits CRC field. The 8-bits CRC field is for validate the integrity of the length of the header of MPDU. The receiver will parse the length field to de-capsulate the following MPDU. In case of corrupt delimiter, i.e., invalid length or CRC, receiver will discard the MPDU and move toward next MPDU. Here, multiple MPDU can be transmitted in a single channel access. Bandwidth hungry and delay sensitive applications which generates burst of small packets will not suffer more as in a single channel access opportunity, multiple frames will be transmitted. A single PPDU is transmitted.

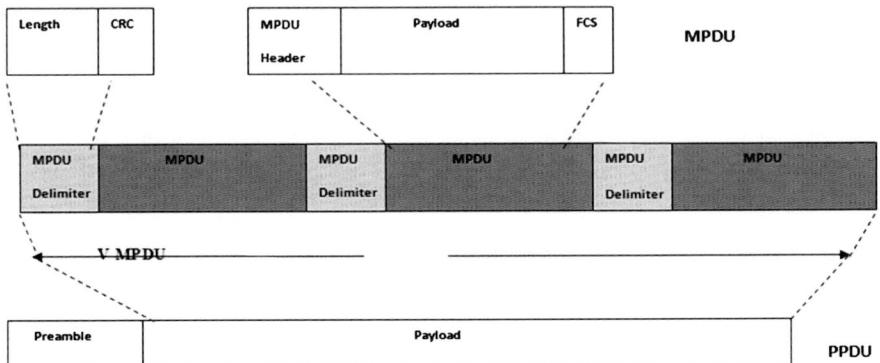

**Fig. 2** V-MPDU encapsulation

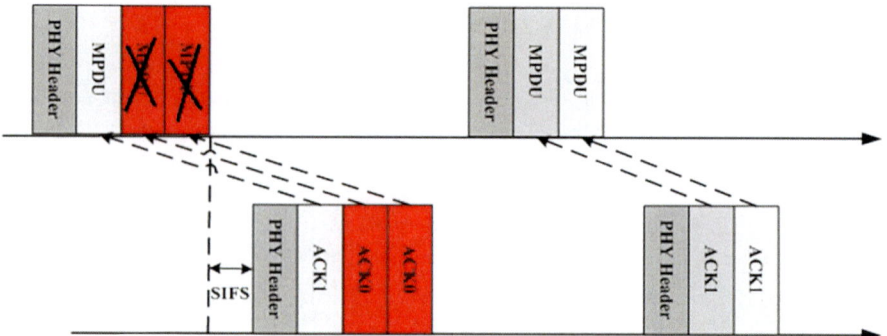

**Fig. 3**  Block acknowledgment for V-MPDU

## 2.2 Block Acknowledgment

IEEE 802.11e introduced the block acknowledgment (BA) mechanism [9], i.e., a block of data is acknowledged with a single BA instead of sending acknowledgments for each frame. Block acknowledgment [3, 14] feature handles all the acknowledgments of the individual frames produced by V-MPDU aggregation into a single frame returned by the recipient to the sender. This allows to implementation of selective re-transmission only for not acknowledged frames. In high erroneous environment rates, this selective re-transmission mechanism can provide some improvement in the effective throughput. Figure 3 depicts the Block Acknowledgment mechanism.

## 2.3 Conditions for the V-MPDU

The purpose of V-MPDU is to combine multiple MPDUs under a single PHY header. Although there is no TID matching condition for the formation of V-MPDU but all the MPDUs in V-MPDU must have the same destination address. No waiting time is required for the formation of V-MPDU as it will take the available packets from the transmission queue. Padding bytes can be attached at the end of MPDU to make the size of MPDU compatible. The default channel access mechanism of IEEE 802.11p is used for the channel access. Hence, a packet can be transmitted under two conditions

- A V-MPDU will be transmitted as a node get the channel access, i.e., it will collect all MPDUs from queue, make a V-MPDU and transmits,
- A V-MPDU will be transmitted as the maximum size of V-MPDU is achieved.
- For emergency applications the V-MPDU is optional.

**Table 1** General simulation parameter

| Operating system | Linux distribution fedora core 9 |
|---|---|
| NS-2 versions | NS-2.34 for IEEE 802.11p |
| | NS-2.34 for IEEE 802.11p with V-MPDU |
| Radio-propagation model | Two-ray-ground |
| Bandwidth | 27 Mbps |
| Traffic flows | Constant-bit rate (CBR) |
| CBR flow rates (Mb/s) | Varies from 1 to 25 Mbps |
| Transport layer protocol | UDP |
| Number of nodes | Varies from 1 to 60 |

# 3 Simulation and Analysis

This section describes simulation setup used for the comparative analysis of IEEE 802.11p and IEEE 802.11p with V-MPDU. For the simulation, we have selected NS-2 which is an open source network simulator. For IEEE 802.11p simulation, we have used NS-2.34 which has a built-in support for IEEE 802.11p. For IEEE 802.11p V-MPDU simulation, we have used NS-2.34 and modifies the A-MPDU code to V-MPDU.

Table 1 describes parameters of simulation setup. Simulations are done in Linux distribution Fedora Core 9. Bandwidth is assumed as 27 Mbps for both IEEE 802.11p and IEEE 802.11p with V-MPDU. Constant-bit rate (CBR) traffic flows are used with UDP. Table 2 shows the parameters settings for IEEE 802.11p and IEEE 802.11 with V-MPDU.

For analysis, we are using three metrics: goodput, average end-to-end delay, and jitter. These three parameters are selected because of their significant role in estimating efficiency of real time and bandwidth hungry applications.

For comparative analysis, we have selected the following simulation scenarios

- Application traffic load
- Node density.

## 3.1 Application Traffic Load

In this scenario, we compare the performance of IEEE 802.11p and IEEE 802.11 with V-MPDU for bandwidth hungry applications. Table 3 gives the requirements for few bandwidth hungry applications. For evaluation, we examine effect on goodput and delay by varying traffic loads. In this scenario, same bandwidth (27 Mbps) is set for both IEEE 802.11p and IEEE 802.11p with V-MPDU, which is the maximum bandwidth value provided by IEEE 802.11p.

**Table 2** Simulation parameters setting for MAC

| Slot time | 13 ns |
|---|---|
| SIFS interval | 32 ns |
| Contention window min (CWMin) | 15 slots |
| Contention window max (CWMax) | 1,023 slots |

**Table 3** Bandwidth hungry applications

| Applications | Bandwidth (Mbps) | Delay (ms) |
|---|---|---|
| High-definition telepresence | 24 | 50 |
| Telemedicine and remote surgery | 10 | 1 |
| Video instant messaging and video presence | 10 | 4 |
| High-definition television | 8 | |
| Real-time data backup | 2 | 10 |

CBR traffic flows is used in the simulation with packet size of 1,000 bytes which is kept constant. There are two other CBR flows of 500 Kbps which acted as background traffic. 15-nodes are used in the scenario, which were moving at 100 km/h. CBR traffic rate is varied to check its effect on goodput and packet delay at receiving node.

Bandwidth is considered as 27 Mbps for both IEEE 802.11p and IEEE 802.11p with VMPDU. Results are shown in Fig. 4 Goodput increases with the increase in traffic load, which were varied as 1, 5, 10, 15, 20, and 25 Mbps. Initially, both IEEE 802.11p and 802.11p with V-MPDU show increase in goodput. A constant goodput of 11 Mbps is observed in case of 802.11p for traffic loads of 15, 20, and 25 Mbps. The reason for this constant goodput is the bandwidth constraint of IEEE 802.11p. Therefore, it does not increase by increasing traffic load. Whereas, IEEE 802.11p with V-MPDU shows goodput of 19 Mbps. IEEE 802.11p with V-MPDU performs well with increasing traffic load. This difference in goodput is due to IEEE 802.11p with V-MPDU MAC layer enhancements (frame aggregation and block acknowledgment). In the aggregation mechanism, different data packets are combined into one frame for transmission. Due to aggregation, overheads of header and backoff mechanisms are reduced because only one frame is transmitted instead of multiple frames. So it increases channel efficiency and results in high gain in goodput. The single block acknowledgment of the aggregated frame makes this aggregation mechanism more efficient. In the simulation, we have used block acknowledgment along with the aggregation mechanism.

Figure 5 shows end-to-end delay results for IEEE 802.11p and IEEE 802.11p with V-MPDU. It can be observed that for lower traffic load, delay is between 0 and 1 both for IEEE 802.11p and IEEE 802.11p with V-MPDU. By increasing traffic load, delay of IEEE 802.11p increases and after 10 Mbps traffic load, delay becomes too large. The reason for this significant increase in delay is bandwidth constraint of IEEE 802.11p. When application is generating traffic with high rates, due to bandwidth constraint in IEEE 802.11p, packets in the buffer wait more

**Goodput (kbps) Vs. Application Traffic Load**

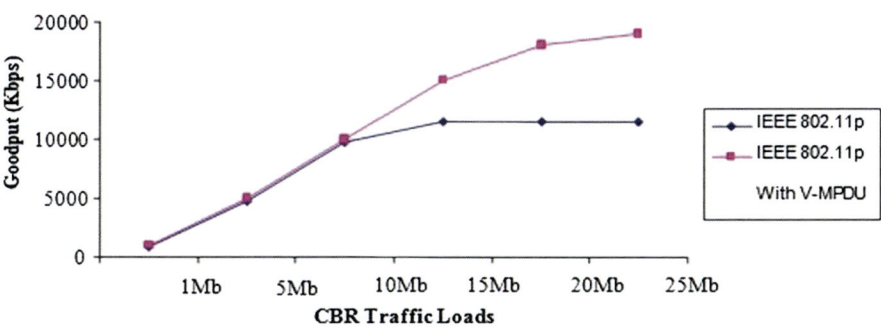

**Fig. 4** Goodput comparison with different application traffic loads

**Delay (ms) with Different Application Rate**

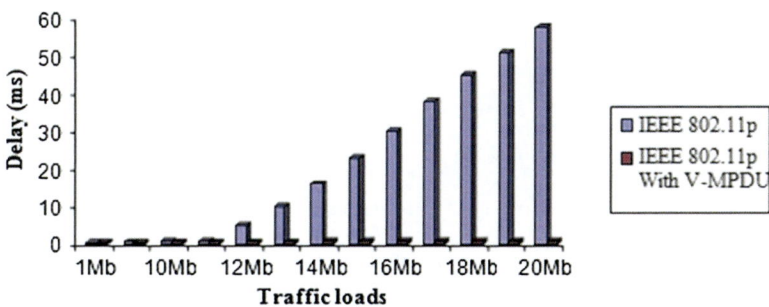

**Fig. 5** Average end-to-end delay with different application traffic loads

which increase the delay. Whereas IEEE 802.11p with V-MPDU shows delay between 0 and 1 ms for all traffic loads. The reason is that, due to aggregation mechanism, overhead of the backoff mechanism and headers is reduced by transmitting a single frame which results in less average end to end delay.

## 4 Conclusion and Future Work

IEEE recommended IEEE 802.11p as the MAC and PHY layer standard for VANET. Our study shows that this standard is not suitable for infotainment applications in the scenario of urban areas where vehicles usually travel with the speed of 40–70 km/h (time critical and bandwidth hungry applications). In this regard, an aggregation mechanism V-MPDU with block acknowledgment as extension of the existing IEEE 802.11p is proposed. V-MPDU aggregation collects frames to be transmitted to a single destination and wraps each frame in a single IEEE 802.11p header. V-MPDU permits each of the aggregated data frames to be individually acknowledged or re-transmitted in case of error. Hence, It improves the channel access mechanism in terms of efficiency as multiple frames are sent in a single transmission which ultimately reduces number of potential collisions and re-transmissions as well. The proposed mechanism shows an improved gain in bandwidth for IEEE 802.11p because of the more simulation scenarios proposed aggregation mechanism. Further, end-to-end delay also reduces. We present the initial simulation results. In future, we will do a detail analysis by adding more scenarios.

# References

1. Amadeo, M., Campolo, C., Molinaro, A.: Enhancing IEEE 802.11 p/wave to provide infotainment applications in vanets. Ad Hoc Netw. **10**(2), 253–269 (2012)
2. Bilstrup, K., Uhlemann, E., Strom, E.G., Bilstrup, U.: Evaluation of the IEEE 802.11 p mac method for vehicle-to-vehicle communication. In: Vehicular Technology Conference, 2008. VTC 2008-Fall. IEEE 68th, pp. 1–5. IEEE (2008)
3. Brown, G.M., Gouda, M.G., Miller, R.E.: Block acknowledgment: redesigning the window protocol. Commun., IEEE Trans. **39**(4), 524–532 (1991)
4. Choi, S., Del Prado, J., Sai Shankar, N., Mangold, S.: IEEE 802.11 e contention-based channel access (edcf) performance evaluation. In: Communications, 2003. ICC'03. IEEE International Conference on vol. 2, pp. 1151–1156. IEEE (2003)
5. Dressler, F., Kargl, F., Ott, J., Tonguz, O.K., Wischhof, L.: Research challenges in intervehicular communication: lessons of the 2010 dagstuhl seminar. Commun. Mag., IEEE **49**(5), 158–164 (2011)
6. Eichler, S.: Performance evaluation of the IEEE 802.11 p wave communication standard. In: Vehicular Technology Conference, 2007. VTC-2007 Fall. 2007 IEEE 66th, pp. 2199–2203 IEEE (2007)
7. Katrin, B., Elisabeth, U., Erik, G.S., Urban, B., et al.: On the ability of the 802.11 p mac method and STDMA to support real-time vehicle-to-vehicle communication. EURASIP J. Wireless Commun. Netw. 2009 (2009)
8. Liu, B., Khorashadi, B., Du, H., Ghosal, D., Chuah, C.N., Zhang, M.: Vgsim: an integrated networking and microscopic vehicular mobility simulation platform. Commun. Mag., IEEE **47**(5), 134–141 (2009)
9. Mangold, S., Choi, S., Hiertz, G.R., Klein, O., Walke, B.: Analysis of IEEE 802.11 e for QOS support in wireless lans. Wireless Commun. IEEE **10**(6), 40–50 (2003)
10. Mangold, S., Choi, S., May, P., Klein, O., Hiertz, G., Stibor, L.: IEEE 802 11 e wireless LAN for quality of service. In: Proc Eur. Wireless **2**, 32–39 (2002)
11. Mohammad, S.A., Rasheed, A., Qayyum, A.: Vanet architectures and protocol stacks: a survey. In: Communication Technologies for Vehicles, pp. 95–105. Springer (2011)
12. Morgan, Y.L.: Notes on DSRC and wave standards suite: Its architecture, design, and characteristics. Commun. Sur. Tutorials, IEEE **12**(4), 504–518 (2010)
13. Murray, T., Cojocari, M., Fu, H.: Measuring the performance of IEEE 802.11 p using ns-2 simulator for vehicular networks. In: Electro/Information Technology, 2008. EIT 2008. IEEE International Conference on, pp. 498–503. IEEE (2008)
14. O'hara, B., Petrick, A.: IEEE 802.11 handbook: a designer's companion. In: IEEE Standards Association (2005)

15. Wang, Y., Ahmed, A., Krishnamachari, B., Psounis, K.: IEEE 802.11 p performance evaluation and protocol enhancement. In: Vehicular Electronics and Safety, 2008. ICVES 2008. IEEE International Conference on, pp. 317–322. IEEE (2008)
16. Wang, Y., Zhang, Y., Jiang, H., Liu, J., Wu, J.: An integrated propagation model for vanet in urban scenario. In: Proceedings of the 6th International Wireless Communications and Mobile Computing Conference, pp. 6–10. ACM (2010)
17. Yang, Y., Bagrodia, R.: Evaluation of vanet-based advanced intelligent transportation systems. In: Proceedings of the 6th ACM international workshop on VehiculAr InterNETworking, pp. 3–12. ACM (2009)

# Part II
# Vanet Security Track

# Invited Paper: VANET Security: Going Beyond Cryptographic-Centric Solutions

Dhavy Gantsou

**Abstract** The need for safety and comfort in vehicular environments has led to a lot of research in vehicular ad-hoc networks (VANET), ranging from vehicle to vehicle (V2V) and vehicle to infrastructure (V2I) communications to security. As regards the latter aspect, mostly the focus has been on resorting to public key infrastructure (PKI) and its related concepts. Moreover, methods used to detect the most feared security attacks are based almost exclusively on the use of physical characteristics as well as PKI concepts or both. However with emerging technologies that enable connecting a vehicle to external resources, VANET is evolving, thus inheriting security threats that are common place in conventional IT systems. Therefore, as it is shown in the paper, VANET security must go beyond cryptographic-centric mechanisms that are commonly used for providing data security in wireless environment.

**Keywords** Security · VANET security · Sybil attack · Honeypot

## 1 Introduction

Vehicular ad-hoc networks (VANET) is a specific kind of mobile ad-hoc networks (MANETs) that provides safety and comfort to drivers in vehicular environments. To achieve this goal, a lot of research efforts have been devoted to VANET in recent years, ranging from vehicle to vehicle (V2V) and vehicle to infrastructure (V2I) communications to security challenges. As regards the latter aspect, mostly the focus has been on resorting to public key infrastructure (PKI) and its concepts

D. Gantsou (✉)
Univ Lille Nord de France, 59000 Lille, France
e-mail: dhavy.gantsou@univ-valenciennes.fr

D. Gantsou
LAMIH-DIM, UMR 8201, 59313 Valenciennes, France

© Springer Science+Business Media Singapore 2015
A. Laouiti et al. (eds.), *Vehicular Ad-hoc Networks for Smart Cities*, Advances in Intelligent Systems and Computing 306, DOI 10.1007/978-981-287-158-9_4

such as certificates, CRL, and digital signature in order to ensure anonymity [1, 2], secure communication [3, 4], and data security including user authentication [5, 6], message authentication [7, 8]. However, with emerging paradigms such as vehicular cloud computing [9, 10] that involves connecting a vehicle to external resources including user devices, corporate networks, internet, and cloud services using cellular networks infrastructure (3G, WiMax, and 4G/LTE), VANET is evolving. The gateway between vehicles and external resources inevitably leads to security attacks that are common place in conventional IT systems.

From the point of view of conventional IT, security strives to preserve the three principles composing the CIA triad made up by confidentiality, integrity, and availability network resource. A resource may be an information, an application, a device or the network itself. However, in VANET, the security concept has different definitions that have led to several security categories [10, 11]. However, they all have three common characteristics:

1. they provide anonymity, user authentication, and data security including message authentication, data encryption to mention few,
2. they are narrowed to security in the IEEE802.11p wireless environment,
3. they are narrowed to essentially resorting to cryptographic mechanisms in order to achieve both the aforementioned and the mitigation of some feared attacks such as the Sybil attack

Yet with the emergence of new paradigms such as cloud, VANET are likely to evolve. However, the shift to vehicular cloud exposes the VANET to the same threats as conventional IT that cannot be addressed using current VANET security approaches.

While it may be simply impossible to prevent attackers from attempting to carry out an attack in VANETs, it might be possible to collect and to analyze data they generate, in order to lessen both the number and the effects of the attacks. This is the aim of the proactive approach relying on Honeypot that is presented in [12, 13]. In general, a honeypot is a computing resource [12], whose sole task is to be probed, attacked, compromised, used, or accessed in any other unauthorized way. The resource could be essentially of any type: a service, an application, a system or set of systems, or simply just a piece of information/data. The key assumption is that any entity connecting to or attempting to use this resource in any way is by definition suspicious. However, proactive security approach does not guarantee complete prevention.

According to Newsome et al. [14], the conventional approaches to address identity-based attacks use authentication. However, the application of authentication requires reliable key distribution, management, and maintenance mechanisms. It is not always desirable to apply authentication because of its infrastructural, computational, and management overhead. Furthermore, cryptographic methods are susceptible to node compromise, which is a serious concern, because most wireless nodes are easily accessible, allowing their memory to be easily scanned. Consequently, VANET security must be addressed by means of approaches that go beyond the use of cryptographic methods. This concern is the aim of this paper. Although the Sybil attack is not specific to the VANET evolution, we show how it can be detected without using extra hardware to deal with physical characteristics or cryptographic-centric methods requiring a centralized authority as it is commonly done. We show a practical approach that uses honeypots to the pairing of one MAC address with several IP addresses.

The reminder of this paper is organized as follows. Section 2 presents a review of related works on detecting Sybil attack in VANETs. The context of the attack and our experimental methodology are presented in Sect. 3. Finally, Sect. 4 concludes the paper.

## 2 Related Work

In recent years, much interest and consideration have been paid to securing VANET whose importance and societal impact is shown by the proliferation of studies and consortia around the world. A number of approaches to security have been proposed, each attempting to mitigate a specific set of concerns. The specific threat, which is the main focus of this article, is Sybil attack [15].

For a given node interacting with others in the network, performing Sybil attack results in the use of multiple identities in order to behave as if it were a larger number of nodes (instead of just one) [14]. This can be done by theft or by forging identities. Sybil attack may be performed following one of the ways presented in [16]: direct versus indirect communication; fabricated versus stolen identities; and simultaneity.

According to Chen et al. [16], by using a single node to present multiple identities in the network, the Sybil attack can significantly reduce the effectiveness of fault-tolerant schemes such as redundancy mechanisms [17], distributed storage [18], multipath routing [19], and topology maintenance [20]. In [16], it is mentioned that the Sybil attack can defeat the redundancy mechanisms, storage partitions, and routing algorithms by making the mechanisms believe that they are using multiple nodes but are, in fact, using a single Sybil node. Therefore, the identity-based attacks, both spoofing and Sybil attacks, will significantly impact the network performance.

The detection of Sybil attack has been intensively studied in the wireless and sensor networks where three types of defense against Sybil attacks are introduced, including: radio resource testing, identity registration, and position verification [21]. According to Yu et al. [22] as radio resource testing is based on the assumption that a radio cannot send or receive simultaneously on more than one channel, it does not apply to VANETs since a greedy driver may cheaply acquire multiple radios. Identity registration alone cannot prevent Sybil attacks, because of a malicious node may get multiple identities by nontechnical means such as stealing. Further, strict registration causes serious privacy concerns. In position verification, the network verifies the position of each node and ensures that each physical node is bound with only one identity.

A number of position (or distance) verification techniques have been proposed recently. None of them would be suitable for the highly mobile context of vehicular networks. However, they are designed for indoor applications or rely on stationary base stations or specific hardware. References related to these topics, like those related to roadside unit based Sybil node detection in VANETs are presented in [23] where, in order to overcome issues related to these approaches a solution using signal strength distribution analysis to detect Sybil attack is proposed. One may notice that the realization of these method relies on a variety of assumptions. This means that to ensure security, conditions other than security policies must be met. The approach proposed in this paper does not follow the above-mentioned schemes. It does not require central authority or pre-configuration in order to perform detection of Sybil nodes. This is the motivation for designing techniques for tracking IP addresses.

# 3 Context and Experimental Attack Methodology

The approach presented in this study aims at providing a method that enables detecting Sybil attack in VANET communication systems, with a focus on multi-hop V2V communications. The choice of the latter is guided by two reasons: (1) to make inter-vehicle communication independent on connection between vehicles (direct/indirect), (2) to enable access to external resources by car drivers. To that end, a dedicated routing protocol is needed. We choose VANET QoS-OLSR [24]. The latter is one of the multiple extensions of OLSR [25]. It is based on ant colony optimization that is able to:

- Extend the network lifetime and maintain the QoS requirements by introducing a QoS-based clustering algorithm that considers the mobility metrics.
- Enhance the end-to-end delay and the Packet Delivery Ratio by selecting the multi points relay (MPR) nodes using ACO.
- Prevent the cheating during the MPR nodes selection using an encryption algorithm.
- Reduce the overhead by introducing a MPR recovery algorithm that is able to select alternative MPRs in case of link failures.

The basic idea in OLSR is to elect a cluster-head for each group of neighbor nodes and divide the network into clusters. These heads then select a set of specialized nodes called MPRs. The function of the MPR nodes is to reduce the overhead of flooding messages by minimizing duplicate transmissions within the same zone. Thanks to their role, MPR will also act as Sybil node detector.

## 3.1 On Sybil Attack Detection in OLSR Based V2V Communication

The first step toward experimental detection of Sybil attack is the creation of the VANET. Network nodes are laptops running:

- Ubuntu 13.04
- BackTrack 5

- Centos 6.5
- Windows 2K SP2
- OS X 10.9

Following software were also used:

- OLSRD 0.6.6.1
- ARPWatch

Finally, in order to collect information that are needed for detecting malicious nodes that are capable to lunch Sybil attack, we used Honeyd 1.5-C to implement a low interaction honeypot as mentioned in [15].

## 3.2 Sybill Attack Implementation

In order to achieve this goal, one needs a malicious node that is capable of having multiple identities. We focus on forged identities. To that end, we proposed an algorithm that enables both creating, and associating multiple IP addresses to the physical interface (MAC-address) of the node. Unlike approaches mentioned in the literature, there is no pre-registration of both MAC, and IP addresses of regular nodes. They are anonymously recorded by ARP protocol as usual. Consequently, the privacy is preserved.

Forged IP addresses are then detected by means of the correlation of MAC and IP addresses. Figure 1 shows that the IP addresses 10.42.43.77 and 10.42.43.61 are associated to the same physical interface e0:ca:94:0b:01:aa.

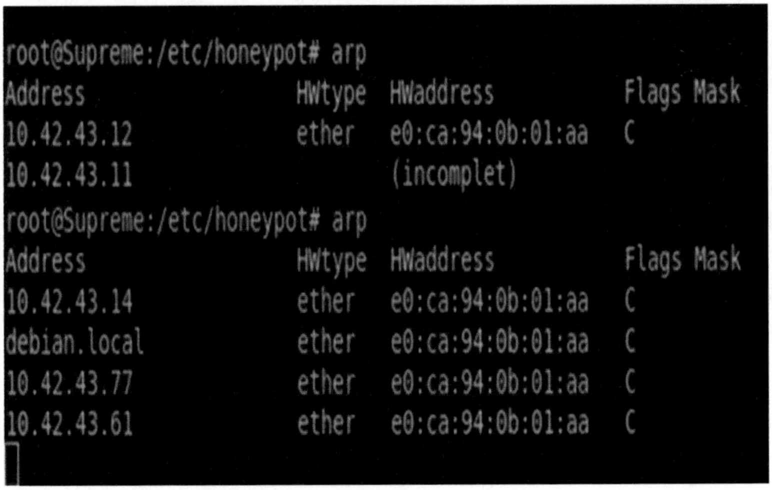

```
root@Supreme:/etc/honeypot# arp
Address                    HWtype   HWaddress              Flags Mask
10.42.43.12                ether    e0:ca:94:0b:01:aa      C
10.42.43.11                         (incomplet)
root@Supreme:/etc/honeypot# arp
Address                    HWtype   HWaddress              Flags Mask
10.42.43.14                ether    e0:ca:94:0b:01:aa      C
debian.local               ether    e0:ca:94:0b:01:aa      C
10.42.43.77                ether    e0:ca:94:0b:01:aa      C
10.42.43.61                ether    e0:ca:94:0b:01:aa      C
```

**Fig. 1** IP-MAC addresses correlation

```
root@Supreme:/home/supreme/Téléchargements# cat /var/lib/arpwatch/eth1.dat
e0:ca:94:0b:01:aa        10.42.43.183      1388844717              eth1
00:25:56:59:a8:59        10.42.43.1        1388844806      Supreme eth1
e0:ca:94:0b:01:aa        10.42.43.216      1388844717              eth1
e0:ca:94:0b:01:aa        10.42.43.24       1388844717              eth1
e0:ca:94:0b:01:aa        10.42.43.135      1388844717              eth1
e0:ca:94:0b:01:aa        10.42.43.69       1388844717              eth1
e0:ca:94:0b:01:aa        10.42.43.240      1388844717              eth1
e0:ca:94:0b:01:aa        10.42.43.101      1388844717              eth1
e0:ca:94:0b:01:aa        10.42.43.230      1388844717              eth1
e0:ca:94:0b:01:aa        10.42.43.196      1388844717              eth1
e0:ca:94:0b:01:aa        10.42.43.149      1388844717      debian  eth1
e0:ca:94:0b:01:aa        10.42.43.12       1388844806              eth1
root@Supreme:/home/supreme/Téléchargements# ./detect_attack.py
vous etes victime d'une sybill attack
10
root@Supreme:/home/supreme/Téléchargements#
```

**Fig. 2** Network victim of 10 attacks

Besides the detection of Sybil node, it may be interesting to monitor whether or not the network is being attacked, and how many nodes are attacking the network. Figure 2 shows a visualization of a network that is victim of 10 attacks.

# 4 Conclusion

We proposed a practical way of detecting Sybil malicious nodes forging identities that can be used for Sybil attack, focusing on multi-hop V2V communication protocols. The approach relies on honeypot and dos not make use of cryptographic methods. It does not require centralized infrastructure or pre-registration of identities. Moreover it can be implemented on commonly used hardware, a good perspective regardless of the smart cities concept. With both the growing size of networks, and the emergence of VANET cloud, it will be necessary to consider the approach in these context. This will be one of the topics of our future investigations.

# References

1. Fischer, L., Aijaz, A., Eckert, C., Vogt, D.: Secure revocable anonymous authenticated inter-vehicle communication (SRAAC). In: 4th Workshop on Embedded Security in Cars (ESCAR 2006), (2006)
2. Fonseca, E., Festag, A., Baldessari, R., Aguiar, R.: Support of anonymity in VANETs-putting pseudonymity into practice. In: Wireless Communications and Networking Conference, IEEE WCNC 2007, pp. 3400–3405 (2007)
3. Sun, X., Lin, X., Ho, P.-H.: Secure vehicular communications based on group signature and ID-based signature scheme. In: Proceedings of International Conference on Communications (ICC'07), Scotland (2007)

4. Chaurasia, B.K., Verma, S., Bhasker, S.M.: Message broadcast in VANETs using group signature. In: Proceedings of the IEEE WCSN'09, pp. 91–96 (2008)
5. Wasef, A., Jiang, Y., Shen, X.: DCS: an efficient distributed-certificate-service scheme for vehicular networks. IEEE Trans. Veh. Technol. **59**, 533–549 (2010)
6. Laberteaux, K.P., Haas, J.J., Hu, Y.C.: Security certificate revocation list distribution for VANET. In: Proceedings of the Fifth ACM Workshop on Vehicular Networks (2008)
7. Wen, H., Ho, P.H., Gong, G.: A novel framework for message authentication in vehicular communication network. In: Proceedings of the IEEE GLOBECOM'09, pp. 1–6 (2009)
8. Zhang, C., Lu, R., Lin, X., Ho, P.H., Shen, X.: An efficient identity-based batch verification scheme for vehicular sensor networks. In: Proceedings of the IEEE INFOCOM '08, pp. 89–824 (2008)
9. Lee, E., Lee, E.-K., Gerla, M.: Vehicular cloud networking: architecture and design principles. IEEE Commun. Mag. pp. 148–155 (2014)
10. Yan, G., Wen, D., Olariu, S., Weigle, M.C.: Security challenges in vehicular cloud computing. IEEE Trans. Intell. Transp. Syst. **14**(1), 284–294 (2013)
11. Goudarzi, S. et al.: A systematic review of security in vehicular ad hoc network. In: Proceedings of 2nd Symposium on Wireless Sensors and Cellular Networks (WSCN'13), Jeddah, Saudi Arabia, 13–16 Dec 2013
12. Pitzner, L.: Honeypots: Tracking Hackers. Addison-Wesley, Boston (2000)
13. Gantsou, D., Sondi, P.: Toward a honeypot solution for proactive security in vehicular ad hoc networks. In: Future Information Technology. Lecture Notes in Electrical Engineering, vol. 276, pp. 145–150 (2014)
14. Newsome, J., Shi, E., Song, D., Perrig, A.: The Sybil attack in sensor networks: analysis and defenses. In: Third International Symposium on Information Processing in Sensor Networks, IPSN 2004, pp. 259–268 (2004)
15. Douceur, J.R.: The Sybil attack. In: Proceedings of 1st IPTPS, Mar 2002, pp. 251–260 (2002)
16. Chen, Y., Yang, J., Trappe, W., Martin, R.P.: Detecting and localizing identity-based attacks in wireless and sensor networks. IEEE Trans. Veh. Technol. **59**(5), 2418–2434 (2010)
17. Zhang, Q., Wang, P., Reeves, D., Ning, P.: Defending against Sybil attacks in sensor networks. In: Proceedings of 25th IEEE ICDCSW, June 2005, pp. 185–191 (2005)
18. Castro, M., Liskov, B.: Practical Byzantine fault tolerance. In: Proceedings of OSDI, pp. 173–186 (1999)
19. Banerjea, A.: A taxonomy of dispersity routing schemes for fault-tolerant real-time channels. In: Proceedings of ECMAST 1999, vol. 26, pp. 129–148 (1999)
20. Chen, B., Jamieson, K., Balakrishnan, H., Morris, R.: Span: an energy efficient coordination algorithm for topology maintenance in ad hoc wireless networks. ACM Wirel. Netw. J. **8**(5), 481–494 (2002)
21. Golle, P., Greene, D., Staddon, J.: Detecting and correcting malicious data in VANETs. In: Proceedings of ACM International Workshop on Vehicular Ad Hoc Networks (VANET 2004), pp. 29–37 (2004)
22. Yu, B., Xu, C.-Z., Xiao, B.: Detecting Sybil attacks in VANETs. J. Parallel Distrib. Comput. (2013). doi:10.1016/j.jpdc.2013.02.001J
23. Park, S., Aslam, B., Turgut, D., Zou, C.C.: Defense against Sybil attack in vehicular ad hoc network based on roadside unit support. In: Proceedings of Military Communications Conference (2009)
24. Wahab, O.A., Otrok, H., Mourad, A.: VANET QoS-OLSR: QoS-based clustering protocol for vehicular ad hoc networks. Comput. Commun. **36**, 1422–1435 (2013)
25. CLausen, T., Jacquet, P. (eds.): Projet Hipercom, INRIA. Oct 2003

# Attacks on Security Goals (Confidentiality, Integrity, Availability) in VANET: A Survey

Irshad Ahmed Sumra, Halabi Bin Hasbullah and Jamalul-lail Bin AbManan

**Abstract** In recent years, the VANET has received a greater attention among researchers in academia and industry due to its potential safety application and nonsafety application. Malicious users are one of the types of attackers in VANET and create the security problems. Confidentiality, integrity and availability (CIA) are major components of security goals. The increasing research interest, potential applications, and security problem in VANET lead to the needs to review the attacks on security goals. In this paper, the aim is to present the survey of attacks on security goals and to describe in details the nature of attacks and the behaviour of attackers through different scenarios in the network. The paper also provides a better understanding of security goals and finally it provides an analysis and classifies the attacks on the basis of security goals into different threat levels that can help in the implementation of VANET in real life.

**Keywords** Vehicular ad-hoc network (VANET) · Security goals · Confidentiality integrity availability (CIA) · Attacks

I.A. Sumra (✉) · H.B. Hasbullah
Computer and Information Sciences Department, Universiti Teknologi PETRONAS,
Bandar Seri Iskandar, 31750 Tronoh, Perak, Malaysia
e-mail: irshad_g00943@utp.edu.my; isomro28@gmail.com

H.B. Hasbullah
e-mail: halabi@petronas.com.my

J.B. AbManan
Advanced Information Security Cluster, MIMOS Berhad, Technology Park Malaysia,
Kuala Lumpur, Malaysia
e-mail: jamalul.lail@mimos.my

© Springer Science+Business Media Singapore 2015                                      51
A. Laouiti et al. (eds.), *Vehicular Ad-hoc Networks for Smart Cities*, Advances
in Intelligent Systems and Computing 306, DOI 10.1007/978-981-287-158-9_5

# 1 Introduction

Road accidents are one of the most serious threats to human lives that can lead to partial or complete disability and results in death. Intelligent transportation system (ITS) is one of the technologies that have improved traffic systems by sending in safety information called road to vehicle communication (RVC) to its users on the highway [1]. Vehicular ad-hoc network (VANET) is a subclass of mobile ad-hoc network (MANET) and is considered a promising approach for future ITS. VANET monitors directly vehicular traffic problems using its safety and nonsafety applications. Basically, VANET comprises of two types of communications: vehicle-to-vehicle (V2V) and vehicle-to-roadside (V2R). The user is the main component of the vehicular network and the objective of this network is to serve the user and provide the right information about the road to the user. Attackers are one of the types of users and those who intentionally create problems for other users of a network by launching different types of attacks (passive or active) [2]. In a vehicular network, they become more prominent because they can potentially change life critical message or broadcast a wrong message to other users of the network. Security is an important factor and confidentiality, integrity, availability (CIA) are the major security requirements in vehicular network [3, 4]. It is required that all components [users, vehicle, and road side unit (RSU)] of vehicular network should be secure and work properly to serve the users and achieve the security goals. Figure 1 shows the relationship of attacker with security goals (CIA) in VANET and Fig. 2 shows all the possible attacks related to the security goals in VANET. Detailed descriptions of all the attacks are given in the upcoming sections with some scenarios.

The rest of paper is divided into five sections; Sections 2–4 discuss in detail the basic concept of security goals (CIA) and all possible attacks in VANET. Section 5 enlightens the conclusion of this survey work.

# 2 Attacks on Confidentiality in VANET

Confidentiality [5, 6] is an important security requirement in vehicular communication. A vehicle sends and receives safety and nonsafety messages from V2V and vehicle to infrastructure (V2I). The content of the message should be secure and

**Fig. 1** Attacker with security goals (CIA) in VANET

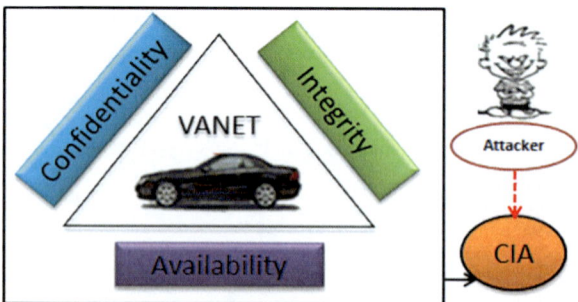

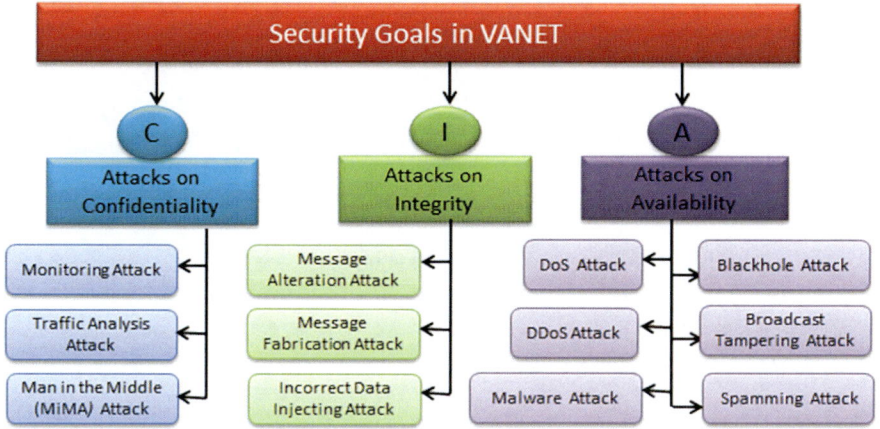

**Fig. 2** All possible attacks on security goals (CIA) in VANET

should not be accessible to nonauthenticated users (attackers). Another aspect of confidentiality is to make the analysis of the traffic flow from V2V or from vehicle to RSU communication. This is a passive-type attack, in which attackers are just monitors the communication between vehicles and gathers require information. All possible attacks related to confidentiality are given below through different scenarios [7, 8].

## 2.1 Monitoring Attack

The attacker in a monitoring attack [9] simply monitors the whole network, listening to the entire communication taking place in V2V and vehicle to roadside unit (V2R). When he/she hears any data that is pertinent to his/her needs then he/she relays this data to the person of interest. One example, in the case of a police operation, the police have planned an operation against a particular criminal to take place in a specified area. To carry out the operation, the police must communicate with each other to pass on the details, such as the exact location and time that the operation is planned. Attackers listen to all of the communication and inform the criminals about the impending police operation. Figure 3 explains the whole scenario where attacker X simply monitors other vehicles communications.

## 2.2 Traffic Analysis Attack

The traffic analysis attack [10] is a serious level threat to user privacy in vehicular communication. A traffic analysis attack is against the anonymity of the communication between (V2V) and vehicle to a roadside unit (V2R). In this attack, the

**Fig. 3** Monitoring attack in
V2V comm

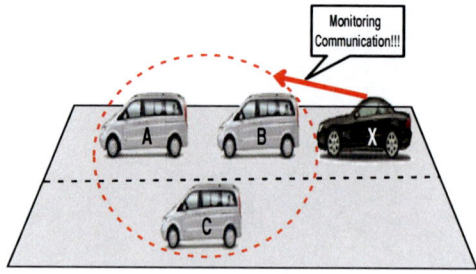

**Fig. 4** Traffic analysis attack
in V2V and V2R

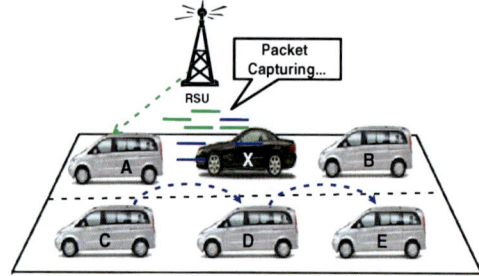

attacker defines some goal and achieves the goal through capturing different types
of traffic information packets. This includes the location of the user, the vehicle
ID, the traveling route of the user or some other of the user's traffic information;
the attacker needs this information to use for its attacks. In Fig. 4, the attacker
captures packets from V2V communication and vehicle for RSU communication.
Attacker X analyses these captured packets and uses them to extract the required
information.

## 2.3 Man in the Middle Attack

Man in the middle (MiMA) attack [11] is a common attack on the communication
that takes place among users. The attacker is usually situated between a minimum
of two persons. The attacker is actively eavesdrops and connects independently to
the vehicle of the victim.

In a MiMA attack, there are two actions that can be carried out by the attacker.

- Eavesdropping on communication between vehicles

  In this situation, both the sender and the receiver think that they are in direct
  communication with each other but in reality, their communication is being
  overheard by an attacker.

  The details of the MiMA attack can be seen in Fig. 5. In this scenario, communication is taking place between vehicle C and vehicle D, and vehicle X is

**Fig. 5** Man in the middle
(MiMA) attack

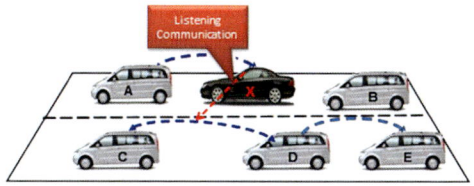

an attacker. Both vehicles C and D think that their communication is not only direct but also secure. The attacker simply eavesdrops on their communication and then uses the data gained for his/her own needs.

# 3 Attacks on Integrity in VANET

In vehicular network, data integrity [5, 6] is one of the most important security goals and it should be maintain while communicating V2V or vehicle to road side unit (V2R). The content of the message should not be altered as it goes from sender to receiver. If the source is authentic user of the network but message contents has been changed then there is no need to check the authenticity of the source user. Message content is very important while communicating in safety and nonsafety applications of vehicular network. All possible attacks related to integrity are given below.

## 3.1 Message Alteration Attack

Malicious Attackers alter the messages, and the false messages are sent to other users. Attackers simply change the content of the safety or nonsafety messages that they have received from other users or from the RUS, then send these alter message to other users in the network [12].

Figure 6 shows the example in which attacker X launches the attack on the safety message. Attacker X receives one warning message *Break down Warning* from vehicle A. So, the attacker changes the content of the message and sends this message *Road is Clear* to vehicle B.

## 3.2 Message Fabrication Attack

In a message fabrication attack [12], attackers broadcast false data in a network. Such kinds of attacks are initiated by greedy drivers. The greedy drivers fabricate messages using broadcast methods and then launch the attack by sending these messages into the network. Fabrication of the messages has two possible forms.

**Fig. 6** Message alteration
attack in V2V

**Fig. 7** Message fabrication
attack

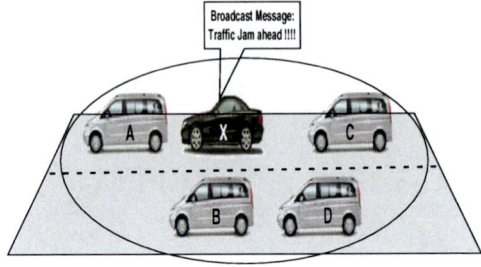

False information about an attacker's ID, speed and location of vehicle is sent to
other vehicles or RSU. Another possibility is that the attacker will present himself/
herself as an emergency vehicle, so that he/she can drive at a faster speed. Figure 7
explains the condition in which attacker X broadcasts the wrong message into the
network.

## 3.3 Incorrect Data Injecting Attack

In this attack, attacker X controls communication by injecting alternative data into
the original message (blue lines) from vehicle C to vehicle D. Figure 8 explains
the behaviour of attacker who is injecting bogus data into vehicle to vehicle communication in network.

## 4 Attacks on Availability in VANET

Availability of network is one of the key modules of security goals. The basic
objective of a vehicular network is to serve the users through its potential applications and the network should be available every time. But, if the network is not
available for communication then the main goal of the network has become use-
less. All possible attacks related to availability are given below through different
scenarios [7, 8, 13].

**Fig. 8** Attacker injecting
incorrect data

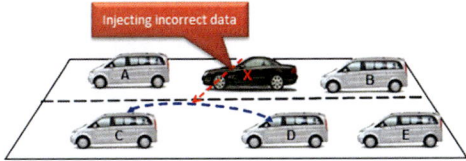

**Fig. 9** DoS attack in V2V
and V2R comm

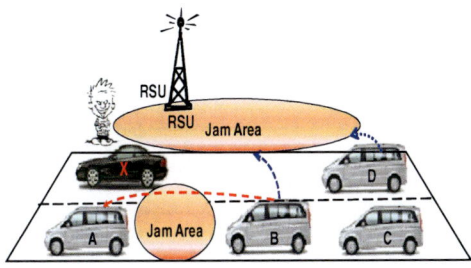

## 4.1 Denial of Service Attack

Denial of service (DoS) [14] attack is one of the key attacks in relation to the availability of the network. Channel jamming in wireless environments is also a part of attack and the objective of the attacker is to prevent the authentic vehicles from accessing the network services. The attack may jam the whole channel or may create some problems directly or indirectly to utilize the resources of the networks. The attacker sends high frequency signals and jams the communication channel between the vehicles. These vehicles cannot send or receive safety or non-safety messages on the network. The attacker launches the attacks near the RSU and jams the communication channel between the vehicles and the RSU. Figure 9 explains this scenario in which vehicle A could not communicate with the other vehicle B due to a DoS attack.

## 4.2 Distributed Denial of Service Attack

A DoS attack is severe in vehicular environment but a Distributed Denial of Service (DDoS) [14, 15] attack is even more severe because the mechanism of the attack in it is in a distributed manner. In this case, attackers launch attacks from different locations. They may use different time slots for launching attacks. The nature of the attack and time slots may be varied from V2V of that particular attacker. Figure 10 explains the scenario in which a group of attacker's vehicles (C, D, and G) launches a DDoS attack on authentic user vehicle F. After some time, the victim user vehicle F cannot communicate with other vehicles in the network.

**Fig. 10** DDoS attack in V2V comm

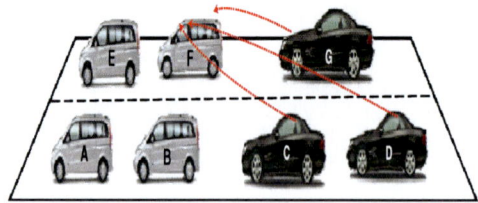

**Fig. 11** Broadcast tampering attack

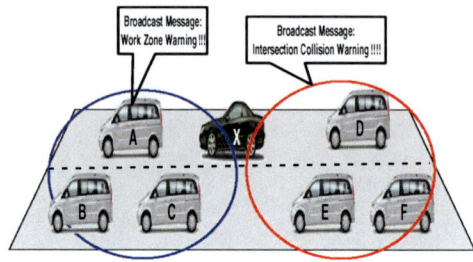

## 4.3 Broadcast Tampering Attack

Safety messages are broadcast in the network and inform other users about current safety conditions of any specific region. In this case, an attacker tampers with the broadcast safety message and possibly injects false safety message. The purpose of this is to cause road accidents or change the flow of traffic on some specific route. Figure 11 shows the behavior of attacker X where the attacker broadcasts two different kinds of messages to two different groups of users.

## 4.4 Malware Attack

A vehicle has its own software and application unit (AU) which performs its own task and communicates with other users as well as the RSU. There is some possibility to enter a virus and worm into the vehicle and disturb the operation of the network. Figure 12 explains the scenario in which a user sends a request to the RSU for software updates. The RSU is already controlled by an attacker, so the attacker downloads the malicious software into the vehicle which made the request. Now this software creates problems for the users.

## 4.5 Spamming Attack

In this situation, the sole purpose of the attacker is to increase the latency of the transmission and use the bandwidth of the network. So, no service is available to

**Fig. 12** Malware attack in
V2R comm

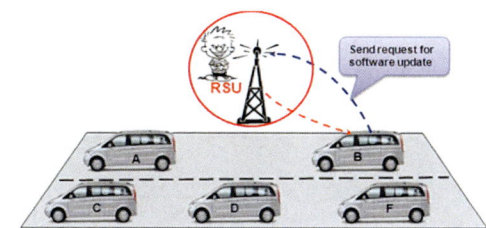

**Fig. 13** Broadcast spam
message attack

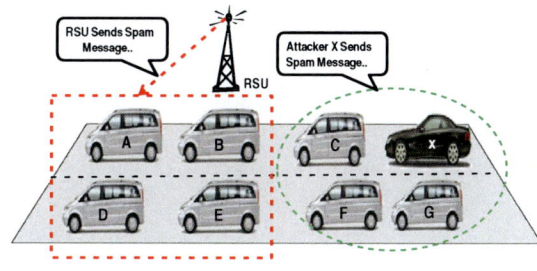

**Fig. 14** Blackhole attack in
V2V comm

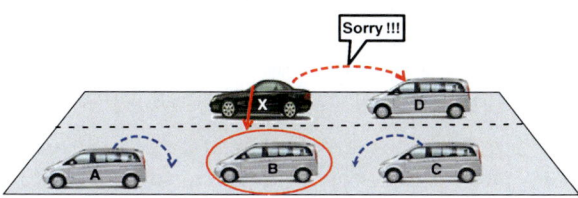

other users of network and this is achieved by sending spam messages through the network. Figure 13 shows the situation when attacker X broadcasts spam messages to a particular group of users. RSU also send spam messages, which are most often just advertisements, to the group of users.

## 4.6 Blackhole Attack

Blackhole attack is a different kind of attack, and there are following two possible cases in vehicular network.

- When any new user wants to start communication with other users or simply participate in a network, then other users simply refuse it. In Fig. 14, user D wants to start communication with user X, but user X refuses it and simply sends a reply with "SORRY." So now user D tries to communicate with any other user of that network.
- One user starts communication with other users of the network and it is suddenly dropped out of the communication. Figure 14 explains the situation in which user B communicating with user A and user C. User B plays the role of

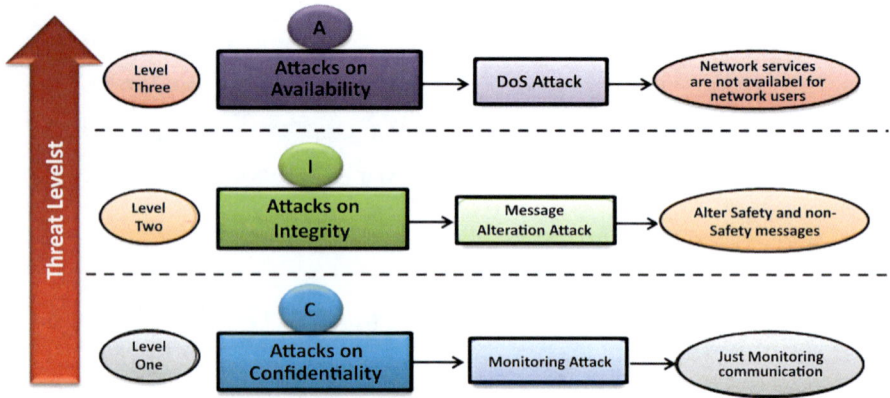

**Fig. 15** Security goals (CIA) with threat levels in VANET

router and sends and receives messages from user A to user C. Attacker X drops the communication of user B and the other neighboring vehicles are disturbed because this vehicle was performing the routing task and many vehicles were connected through it as the router client. In this way, all possible links are down due to the dropping of the link with this intermediate vehicle.

Figure 15 describes different threat levels of security goals in a vehicular network. Every module in a security goal has an important role and we have classified the security goals into three levels (threat levels) in VANET. An availability requirement in network has more priority and more threat level as compared to integrity and confidentiality. Availability is considered as the most important module and is placed at the maximum (third) level, which is related to the availability of network resources. For example, the DoS attack directly affects the network resources and the authentic users of the network will not be able to communicate due to the unavailability of network resources. If we analyze the attacks on integrity, attacker's objective is to alter the safety and nonsafety messages. Due to this attack, the communication of a user can be compromised; however, the user is still able to use the network services. Hence, the attacks on integrity are classified as level two attacks. Moreover, the confidentiality is considered at level one (lowest threat level) and most attacks on it are passive attacks (monitoring attack) in a network that has no threat to the functionality of network.

## 5 Conclusion

Based on the literature survey, it is understood that attackers launch different types of attacks while communication is in progress in the network. These attacks break the security goals like CIA in the VANET environment. These goals have equal importance to serve the users but the availability leads to high priority. It is observed that the attacks related to the availability have more threat level as compare to the integrity and confidently. The accomplishment of the security goals by addressing the nature of attacks and the behaviors of attackers will help to successfully implement the VANET in real environment.

# References

1. Sumra, I.A., Hasbullah, H., bin Ab Manan, J.L.: Classification of Traffic system in intelligent transportation system (ITS). Mod. Traffic Transp. Eng. Res. **2**(4), 1–10 (2013)
2. Raya, M., Hubaux, J.P.: The security of vehicular ad hoc networks. In: Proceedings of the 3rd ACM workshop on Security of Ad hoc and Sensor Networks (2005)
3. Papadimitratos, P., Gligor, V., Hubaux, J.-P.: Securing vehicular communications—assumptions, requirements, and principles. In: Proceedings of the Workshop on Embedded Security in Cars (ESCAR) (2006)
4. Qian, Y., Moayeri, N.: Design of secure and application-oriented VANETs. In: Vehicular Technology Conference, pp. 2794–2799. VTC Spring 2008, IEEE (2008)
5. Plossl, K., Nowey, T., Mletzko, C.: Towards a security architecture for vehicular ad hoc networks. In: The First International Conference on Availability, Reliability and Security, ARES, p. 8 (2006)
6. Engoulou, R.G., Bellaïche, M., Pierre, S., Quintero, A.: VANET security surveys. Comput. Commun. **44**, 1–13 (2014) ISSN 0140-3664
7. de Fuentes, J.M., González-Tablas, A.I., Ribagorda, A.: Overview of security issues in vehicular ad-hoc networks. In: Handbook of Research on Mobility and Computing: Evolving Technologies and Ubiquitous Impacts IGI Global (2011)
8. Zeadally, S., Hunt, R., Chen, Y-Sh, Irwin, A., Hassan, A.: Vehicular ad hoc networks (VANETS): status, results, and challenges. Telecommun. Syst. **9**, 1–25 (2010)
9. Raya, M., Hubaux, J.P.: Securing vehicular ad hoc networks. J. Comput. Secur. **15**, 39–68 (2007)
10. Isaac, J.T., Zeadally, S., Camara, J.S.: Security attacks and solutions for vehicular ad hoc networks. IET Commun. **4**(7), 894, 903 (2010)
11. Sumra, I.A., Hasbullah, H., bin Ab Manan, J.-L.: Behavior of attacker and some new possible attacks in vehicular ad hoc network (VANET). In: 3rd International Congress on Ultra Modern Telecommunications and Control Systems and Workshops (ICUMT), pp. 1, 8, 5–7 Oct 2011
12. Parno, B., Perrig, A.: Challenges in securing vehicular networks. In: Hot Topics in Networks (HotNets-IV) (2005)
13. Sumra, I.A., Hasbullah, H., Ab Manan, J.-L.: Effects of attackers and attacks on availability requirement in vehicular network: a survey. In: International Conference on Computer and Information Sciences (ICCOINS2014), Malaysia, 3–5 June 2014
14. Hasbullah, H., Sumra, I.A., Ab Manan, J.-L.: Denial of service (DOS) attack and its possible solution in VANET. WASET (2010)
15. Biswas, S., Misic, J., Misic, V.: DDoS attack on WAVE-enabled VANET through synchronization. In: Global Communications Conference (GLOBECOM), 2012 IEEE, pp. 1079, 1084, 3–7 Dec 2012

# Part III
# Data Dissemination in Vanet Track

# Performance Evaluation of EAEP and DV-CAST Broadcasting Protocols in VANET

**Warid Islam and Rajesh Palit**

**Abstract** The use of vehicular ad-hoc network (VANET) has been increasing immensely over the past few years. VANET has played a very important role in safety issues on roads. There are many routing protocols in VANET, out of which broadcasting protocol is used more frequently for sharing, traffic, weather and emergency, road conditions among vehicles and delivering advertisements and announcements. Several types of broadcasting protocols have been proposed, out of which DV-CAST protocol and Edge Aware Epidemic Protocol (EAEP) are two such types that have come into force recently. It is said that these two protocols can be of great use later but their performances are yet to be explored in different scenarios. So in this paper the performance of DV-CAST protocol and EAEP are discussed, in well-connected and totally disconnected highway and city scenarios respectively, and a conclusion was reached about which protocol works better in different scenarios. The evaluation includes the measurements of the parameters such as packet drop and throughput. NCTUns 6.0 was used as the network simulator. NCTUns is a high fidelity and extensible network simulator and emulator which helps to simulate complex scenarios. The challenges and future perspectives of broadcasting protocols are also discussed in this paper. This work helps the researchers, who are currently working on other routing protocols in VANET, to come to a decision about the best routing protocols that are to be used in different scenarios.

**Keywords** Vehicular ad-hoc network · Broadcasting protocols · Performance evaluation

W. Islam · R. Palit (✉)
North South University, Dhaka, Bangladesh
e-mail: rpalit@northsouth.edu

© Springer Science+Business Media Singapore 2015      65
A. Laouiti et al. (eds.), *Vehicular Ad-hoc Networks for Smart Cities*, Advances
in Intelligent Systems and Computing 306, DOI 10.1007/978-981-287-158-9_6

# 1 Introduction

The demand for wireless communication has been increasing every day. Due to the need of wireless devices, more researches are done without the interference of a centralized network or infrastructure. The networks with the absence of any centralized or pre-established infrastructure are called Ad-hoc networks. Vehicular ad-hoc networks (VANET) is a subclass of mobile ad-hoc networks (MANET), in which the individual vehicles are considered as nodes in the network.

Unlike MANET, vehicles in VANET move on predefined roads, vehicles velocity depends on the speed signs, and in addition these vehicles also have to follow traffic signs and traffic signals. VANET provides safety and traffic management. Vehicles have to notify other vehicles about hazardous road conditions, traffic jamming, and road stops. VANET has some other applications as well, such as driver assistance, information about latest weather conditions, automatic parking, entertainment, etc.

Routing protocols in VANET are generally classified into unicast protocol, multicast protocol, geocast protocol, and broadcast protocol. In unicast routing protocol, data are transmitted from a single source to a single destination via wireless multihop transmission or carry and forward technique. In the wireless multihop transmission technique, or called as multihop, the intermediate vehicles in a routing path should relay data as soon as possible from source to destination. In the carry-and-forward technique, source vehicle carries data as long as possible to reduce the number of data packets. In geocast multicast routing protocol, the packets are delivered in a specific static geographic region. If the vehicles are located within a specific geographic region, then the vehicles receive the packet. Otherwise, the vehicles drop the packet. In a broadcast protocol, a vehicle disseminates broadcast messages to all other vehicles.

The performance of DV-CAST protocol and *Edge Aware Epidemic Protocol* (EAEP) was evaluated in well-connected and totally disconnected highway and city scenarios, respectively, and a conclusion was reached about which protocol to use in different scenarios. The evaluation includes the measurements of the parameters such as packet drop and throughput. NCTUns 6.0 was used as the network simulator. NCTUns is a high fidelity and extensible network simulator and emulator, which helps us to simulate complex scenarios.

Over the past few years, VANET has evolved as a major network for communications. Its usage has been increasing every day. There are many routing protocols in VANET, out of which broadcasting protocol plays a very important role in disseminating broadcast messages to all other vehicles in a network. There are several classifications of broadcasting protocol. DV-CAST and EAEP protocols are two such types of classifications of broadcasting protocol, which have come into effect recently. These two protocols has been tipped to be used frequently in future circumstances but it has still not been explored widely and the efficiency of these protocols is still not known properly in different scenarios. So well-connected and totally disconnected highway and city scenarios respectively, were considered and a conclusion was reached about which of the above two protocols works better in different scenarios. The performance parameter included packet drop and throughput.

Due to its immense potential, many people are interested to do more research works in VANETs. So we hope this work would help the people who wants to study more about VANETs.

The organization of the paper is as follows. Section 2 outlines a brief description of the broadcasting protocols used in VANET. Section 3 discusses about the comparison of DV-CAST protocol and EAEP in different scenarios. Section 4 contains the performance matrices used in the work. Sections 5 and 6 give a brief outline about the simulation software and parameters used in this paper. Section 7 contains the simulation results and Sect. 8 gives a brief outline about the conclusion of the work and the areas where some further researches can be done in future.

## 2  VANET Broadcasting Protocols

Broadcast routing is frequently used in VANET for sharing, traffic, weather and emergency, road conditions among vehicles and delivering advertisements and announcements. Broadcasting is used when message needs to be disseminated to the vehicle beyond the transmission range, i.e., multi hops are used. Broadcast sends a packet to all nodes in the network, typically using flooding. This ensures the delivery of the packet but bandwidth is wasted and nodes receive duplicates. In VANET, it performs better for a small number of nodes. The various Broadcast routing protocols are BROADCOMM, UMB, V-TRADE, DV-CAST, EAEP, SRB, PBSM, PGB, DECA, and POCA.

BROADCOMM [2] is based on hierarchical structure for highway network. In BROADCOMM, the highway is divided into virtual cells which move like vehicles. The nodes in the highway are divided into two-level hierarchy, the first level includes all the nodes in a cell, the second level is represented by cell reflectors. Cell reflected behaves temporarily as cluster head and handles the emergency messages coming from the same member of the cell or from different members.

In Urban Multihop Broadcast Protocol (UMB) [4], the sender node tries to select the furthest node in the broadcast direction for forwarding and acknowledging the packet without any prior topology information. UMB protocol performs with much success at higher packet loads and vehicle traffic densities. It overcomes hidden node problems and packet collisions during multihop broadcasting.

Vector based tracing detection (V-TRADE) [10] classifies the neighbors into different forwarding groups depending upon position and movement information. For each group, only a small subset of vehicles is selected to rebroadcast the message. It improves the efficiency of bandwidth utilization but also creates overhead problem.

Distributed Vehicular Broadcast protocol (DV-CAST) [11] uses local topology information by using the periodic hello messages for broadcasting the information. Each vehicle uses a flag variable to check whether the packet is redundant or not. This protocol divides the vehicles into three types depending on the local connectivity as well-connected, sparsely connected, totally disconnected neighborhood. In well-connected neighborhood, it uses persistence scheme (weighted p persistence, slotted 1 and p persistence). In sparsely connected neighborhood after receiving the broadcast message, vehicles can immediately rebroadcast with vehicles moving in the same direction. In totally disconnected neighborhood, vehicles are used to store the broadcast message until another vehicle enters into transmission range, otherwise if the time expires it will discard the packet. This protocol causes high control overhead and delay in end to end data transfer.

EAEP [8] is reliable, bandwidth efficient information dissemination based highly dynamic VANET protocol. It reduces control packet overhead by eliminating exchange of additional hello packets for message transfer between different clusters of vehicles and eases cluster maintenance. Each vehicle piggybacks its own geographical position to broadcast messages to eliminate beacon messages. Upon receiving a new rebroadcast message, EAEP uses number of transmission from front nodes and back nodes in a given period of time to calculate the probability for making decision whether nodes will rebroadcast the message or not. But EAEP does not address the intermittent connectivity issue. Specifically, a node does not know whether it has missed any messages to its new neighbors or its neighbors have missed some messages. EAEP overcomes the simple flooding problem but it incurs high delay of data dissemination.

Secure ring broadcasting (SRB) [1] classifies nodes into three groups based on their receiving power as Inner Nodes (close to sending node), Outer Nodes (far away from sending node), Secure Ring Nodes (preferable distance from sending node). It restricts rebroadcasting to only secure ring nodes to minimize the number of retransmissions.

Parameter less broadcasting in static to highly mobile wireless ad-hoc (PBSM) [3] is an adaptive broadcasting protocol that does not require nodes to know about position and movement of their nodes and itself. It uses connected dominating sets (CDS) and neighbor elimination concepts to eliminate redundant broadcasting. It employs two-hop neighbor information obtained by periodic beacons to construct CDS. Each vehicle A maintains two lists of neighboring vehicles: R and NR, containing neighbors that already received and that which did not receive the packet. After a timeout, A rebroadcasts the packet if the list NR is nonempty.

Preferred group broadcast (PGB) [7] is not a reliable broadcasting protocol but it is a solution to prevent broadcast storm problem from route request broadcasting. Each node in PGB will sense the level of signal strength from neighbor broadcasting. The signal strength is used for waiting timeout calculation. Nodes in the edge of circulated broadcast will set shorter waiting timeout. Only node with shortest timeout will rebroadcast the message. PGB can reduce numbers of RREQ broadcasting. But there exists a problem on low density area.

Density-aware reliable broadcasting protocol (DECA) does not require position knowledge. DECA [6] uses only local density information of I-hop neighbors obtained by beaconing. Before broadcasting, a node selects one neighbor, which has the highest local density information to be the next rebroadcast node. Other nodes will randomly set their waiting timeout. If they do not hear anyone rebroadcast the message before the timeout expiration, they will rebroadcast the message.

Position-aware reliable broadcasting protocol (POCA) [5] uses adaptive beacon [9] to get neighbor's position and velocity. When nodes want to broadcast messages, they will select the neighbors in preferred distance to rebroadcast the message. The preferred distance is based on the distance between nodes and selector nodes. The selected node will rebroadcast the message immediately. In case the selected nodes do not rebroadcast the message, other nodes which have set waiting timeout since they received message will do this task instead. The waiting timeout is calculated, depending on the distance between node and precursor node.

## 3 Comparison of EAEP and DV-CAST Protocols

The performance of DV-CAST protocol and EAEP was evaluated in well-connected and totally disconnected highway and city scenarios, respectively. The parameters included number of packet drops and throughput. In DV-CAST protocol, hello messages are used for broadcasting the messages. So this protocol causes high control overhead and a delay in end-to-end data transfer. In EAEP, no hello messages are used for message transfer between different cluster of vehicles. So it reduces flooding problems but incurs high delay of data dissemination.

## 4 Performance Metrices

The performance evaluation of DV-CAST protocol and EAEP in highway and city scenarios were studied in this work. To study their behavior, the packet drop and throughput performance of these two protocols in different scenarios were used.

Throughput: Throughput is the ratio of the total number of received packets at the destination to the total number transmitted packets. Throughput is calculated in bytes/s or data packets per second.

Packet Drop: It shows the total number of data packets that could not reach the destination successfully. The reason for packet drop may arise due to congestion, faulty hardware and queue overflow, etc.

## 5 Simulation Environment

Before commercial deployment of any new technology, realistic testing must be performed. In communication and computer networks research, simulation is the most practical method of evaluation. Simulation allows engineers to test scenarios that might be otherwise difficult or expensive to emulate using real hardware and it allows designers to test new protocols or make changes to existing protocols in a controlled and reproducible environment. Currently, since neither ITS infrastructure nor communications exist, except for small-scale prototypes simulation is the only economically viable and fast way to develop new protocols for ITS.

NCTUns 6.0 was used as the network simulator. The NCTUns network simulator and emulator (NCTUns) is a high-fidelity and extensible network simulator capable of simulating various devices and protocols used in both wired and wireless networks. Its core technology is based on the kernel-reentering simulation methodology invented by Prof. S.Y. Wang at Harvard University in 1999. Due to this novel methodology, NCTUns provides many unique advantages that cannot be easily achieved by traditional network simulator such as OPNET Modeler and ns-2.

## 6 Simulation Parameters

The model for a well-connected highway scenario was constructed by having no obstacles and 50 cars within the transmission range of each other. The transmission range of each car was set at 50 m. The IEEE 802.11(p) MAC protocol was used in all the cars. The simulation time period was 300 s. 1,400 bytes of UDP packets and 15 dBm transmission power were used for node operation. In the graphs below, the simulation time interval is varied in terms of 1 s. The model for the city scenario was made by having obstacles of 20 dBm. All the other parameters remained the same. In DV-CAST protocol, hello messages were transmitted in intervals of 100 ms whereas in EAEP, no hello messages were transmitted. In well-connected scenarios, 50 vehicles were placed close to each other whereas in totally disconnected scenarios, two vehicles were placed far away from each others transmission range.

## 7 Simulation Results

The simulation results of EAEP and DV-CAST protocol in a well-connected and a totally disconnected highway and city scenarios respectively, are shown in this section.

## 7.1 Highway Well-Connected Scenario

**Packet drop** From the graph in Fig. 1a, it can be seen that the maximum number of packet drops in EAEP is about 1,430, where as the maximum number of packet drops in DV-CAST protocol is about 700. So the packet drop in a well-connected highway scenario is higher for EAEP than that of DV-CAST protocol.

**Throughput** From the graph in Fig. 1b, it can be seen that the maximum throughput of DV-CAST protocol is about 725 KB/s where as the maximum throughput of EAEP is about 350 KB/s. So the throughput of DV-CAST protocol is higher than that of EAEP in a well-connected highway scenario. From these two

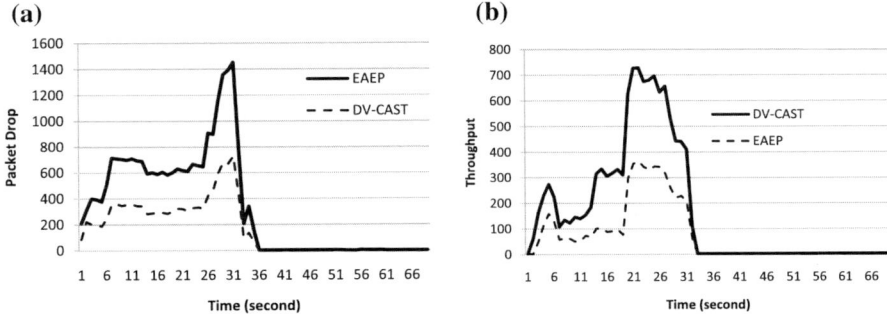

**Fig. 1** Highway well-connected scenario. **a** Packet drop. **b** Throughput

graphs in Fig. 1, it can be seen that the number of packet drops in EAEP is higher than that of DV-CAST protocol, but the throughput of DV-CAST protocol is much higher than that of EAEP. There will be more information loss in EAEP and also it will incur a high delay in reaching its destination. So DV-CAST protocol is more suitable than EAEP in a well-connected highway scenario.

## 7.2 Highway Totally Disconnected Scenario

**Packet drop** In a totally disconnected highway scenario, the packet drop for both DV-CAST protocol and EAEP is zero.

   **Throughput** From the graph in Fig. 2, it can be seen that the maximum throughput of EAEP is about 3.2 KB/s where as the maximum throughput of DV-CAST protocol is about 1.6 KB/s. So the throughput of EAEP is higher than that of DV-CAST protocol in a totally disconnected highway scenario. The packet drop in a totally disconnected highway scenario is zero but the throughput of EAEP is higher than that DV-CAST protocol as seen in Fig. 2. More information will reach

**Fig. 2** Throughput in highway disconnected scenario

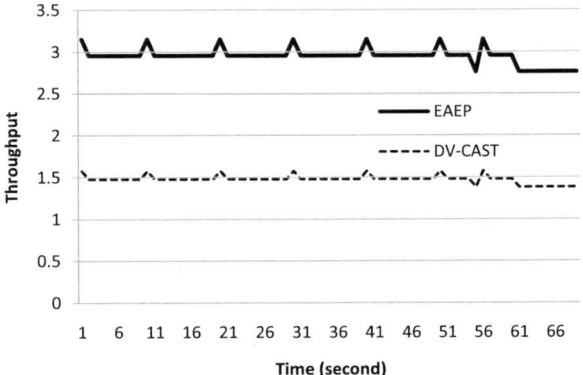

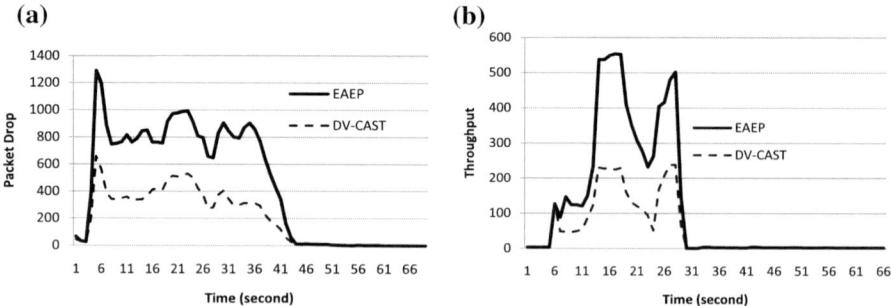

**Fig. 3** City well-connected scenario. **a** Packet drop. **b** Throughput


the destination in EAEP, so EAEP is more suitable than DV-CAST protocol in a totally disconnected highway scenario.


## 7.3 City Well-Connected Scenario

**Packet drop** From the graph in Fig. 3a, it can be seen that the maximum number of packet drops in EAEP is about 1,300 where as the maximum number of packet drops in DV-CAST protocol is about 650. So the packet drop is higher for EAEP than DV-CAST protocol in a well-connected city scenario.

**Throughput** From the graph in Fig. 3b, it can be seen that the maximum throughput of EAEP is about 550 KB/s where as the maximum throughput in DV-CAST protocol is about 225 KB/s. So the throughput of EAEP is higher than that of DV-CAST protocol in a well-connected city scenario. From these two graphs in Fig. 3, it can be seen that the number of packet drops and the throughput of EAEP is higher than that of DV-CAST protocol. Although the loss of information is more in EAEP than that in DV-CAST protocol, but more packets will reach the destination in EAEP in comparison to that in DV-CAST protocol. So EAEP is more suitable to use than DV-CAST protocol in a well-connected city scenario.


## 7.4 City Totally Disconnected Scenario

**Packet Drop** The packet drop is zero for both EAEP and DV-CAST protocol in a totally disconnected city scenario.

**Throughput** From the graph in Fig. 4, it can be seen that the maximum throughput in EAEP is about 3.2 KB/s where as the maximum throughput in DV-CAST protocol 1.55 KB/s. So the throughput in EAEP is higher than DV-CAST protocol in a totally disconnected city scenario. The packet drop for both the protocols is zero in a totally disconnected city scenario but the throughput of EAEP is

**Fig. 4** City totally
disconnected scenario

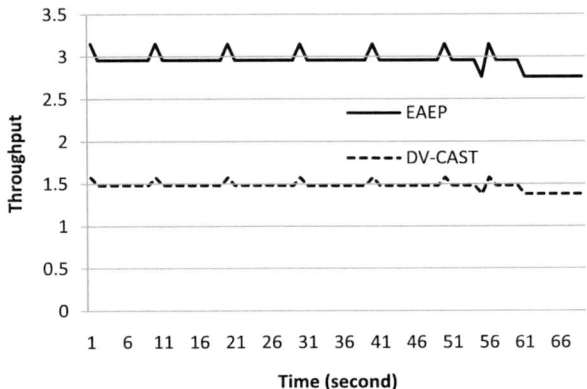

higher than that of DV-CAST protocol. More information will reach the destination in EAEP, so EAEP is more suitable than DV-CAST protocol in a totally disconnected city scenario.

## 7.5 Summary

In a well-connected highway scenario, the packet drop for DV-CAST protocol is lower than that of EAEP. The output throughput of DV-CAST protocol is higher than that of EAEP protocol. So DV-CAST protocol is more suitable to use in a well-connected highway scenario. In a totally disconnected highway scenario, the packet drop for both DV-CAST protocol and EAEP protocol is zero. But the output throughput of EAEP is higher than that of DV-CAST protocol. Therefore, EAEP is more suitable to use than DV-CAST protocol in a totally disconnected highway scenario.

In a well-connected city scenario, the packet drop for DV-CAST protocol is lower than that of EAEP. But the output throughput of EAEP is higher than that of DV-CAST protocol. DV-CAST protocol is more reliable than EAEP but the throughput is less than that of EAEP. So EAEP is more suitable in a well-connected city scenario. In a totally disconnected city scenario, the packet drop for both DV-CAST protocol and EAEP protocol is zero. But the output throughput of EAEP is higher than that of DV-CAST protocol. So EAEP is more suitable to use than DV-CAST protocol in a totally disconnected city scenario.

## 8  Conclusion

The performance of DV-CAST protocol and EAEP was evaluated in a well-connected and a totally disconnected highway and city scenarios, respectively. Number of packet drops and throughput were used as the simulation parameters.

Simulation was performed by using NCTUns 6.0. A highway model was constructed by having no obstacles and a city model was made by having some obstacles. In a well-connected scenario, all the nodes were close to each other where as in a totally disconnected scenario, the nodes were far away from each other. After performing simulation, it was found that DV-CAST protocol is more suitable than EAEP in a well-connected highway scenario where as EAEP is more suitable than DV-CAST protocol in a totally disconnected highway scenario, a well-connected and a totally disconnected city scenario, respectively. More research should be done to see how these protocols works in real world. A suitable broadcast technique should be found that not only has less packet drop but also has high throughput and is also bandwidth efficient, which will work in any environment. The energy distribution of data packets can be studied and the optimum energy level can be found out for different scenarios. The calculation for probability of message transmission in EAEP can also be studied.

# References

1. Bauman, R.: Vehicular ad hoc networks. Master's thesis (2004)
2. Durresi, M., Durresi, A., Barolli, L.: Emergency broadcast protocol for inter-vehicle communications. In: Proceedings of the 11th International Conference on Parallel and Distributed Systems, vol. 2, pp. 402–406 (2005)
3. Khan, A.A., Stojmenovic, I., Zaguia, N.: Parameterless broadcasting in static to highly mobile wireless ad hoc, sensor and actuator networks. In: Proceedings of the 22nd International Conference on Advanced Information Networking and Applications (AINA'08), pp. 620–627. IEEE Computer Society, Washington, DC (2008)
4. Korkmaz, G., Ekici, E., Özgüner, F., Özgüner, Ü.: Urban multi-hop broadcast protocol for inter-vehicle communication systems. In: Proceedings of the 1st ACM International Workshop on Vehicular Ad Hoc Networks, pp. 76–85. ACM (2004)
5. Nakom, K.N., Rojviboonchai, K.: POCA: position-aware reliable broadcasting in vanet. In: Proceedings of the 2nd Asia-Pacific Conference of Information Processing (APCIP2010), pp. 1405–1413 (2010)
6. Nakorn, N.N., Rojviboonchai, K.: DECA: density-aware reliable broadcasting in vehicular ad hoc networks. In: 2010 International Conference on Electrical Engineering/Electronics Computer Telecommunications and Information Technology (ECTI-CON), pp. 598–602. IEEE (2010)
7. Naumov, V., Baumann, R., Gross, T.: An evaluation of inter-vehicle ad hoc networks based on realistic vehicular traces. In: Proceedings of the 7th ACM International Symposium on Mobile Ad Hoc Networking and Computing, pp. 108–119. ACM (2006)
8. Nekovee, M., Bogason, B.B.: Reliable and efficient information dissemination in intermittently connected vehicular ad hoc networks. In: IEEE 65th Vehicular Technology Conference (VTC2007). pp. 2486–2490. IEEE (2007)
9. Park, V.D., Corson, M.S.: A highly adaptive distributed routing algorithm for mobile wireless networks. In: Sixteenth Annual Joint Conference of the IEEE Computer and Communications Societies (INFOCOM'97), Proceedings IEEE, vol. 3, pp. 1405–1413. IEEE (1997)
10. Sun, M.T., Feng, W.C., Lai, T.H., Yamada, K., Okada, H., Fujimura, K.: Gps-based message broadcasting for inter-vehicle communication. In: Proceedings. 2000 International Conference on Parallel Processing, pp. 279–286. IEEE (2000)
11. Tonguz, O., Wisitpongphan, N., Bai, F., Mudalige, P., Sadekar, V.: Broadcasting in vanet. In: 2007 mobile networking for vehicular environments, pp. 7–12. IEEE (2007)

# ZoomOut HELLO: A Novel 1-Hop Broadcast Scheme to Improve Network QoS for VANET on Highways

**Maaz Rehan, Halabi Hasbullah, Waqas Rehan and Omer Chughtai**

**Abstract** On highways, each vehicle uses periodic 1-hop broadcast messages to advertise its position and other information so that vehicles in the vicinity and those which are hops away can find the advertising vehicle for information exchange. During destination discovery, a vehicle issues broadcast messages, which travel hop by hop in search of destination. We argue that since periodic 1-hop messages are sent on regular intervals, so these can be manipulated in such an intelligent way that routing protocols may not need to execute broadcast destination discovery procedure. The continuous periodic HELLO broadcast mechanism can assist the routing protocol in this regard. The proposed ZoomOut HELLO (ZOH) technique introduces a neighbour-cum-forwarding (NF) table. In this scheme, periodic ZOH mechanism will populate the NF table, while the forwarding mechanism will use it. We introduce the concepts of Front and Behind relatives, which are selected out of 1-hop neighbours based on the typed HELLO messages exchanged between 1-hop neighbours. ZOH is a broadcast suppression technique and implicitly provides chain of relatives in the front and behind direction of each vehicle. A routing protocol can therefore use this chain. We have developed analytical model for ad-hoc on-demand distance vector (AODV), preferred group broadcast (PGB), reliable opportunistic broadcast

M. Rehan (✉) · H. Hasbullah
Computer Information Science Department, Universiti Teknologi PETRONAS,
Seri Iskandar, Malaysia
e-mail: maaz_g01945@utp.edu.my; maazrehan@ciitwah.edu.pk

H. Hasbullah
e-mail: halabi@petronas.com.my

M. Rehan · O. Chughtai
COMSATS Institute of Information Technology, WAH CANTT, Pakistan

W. Rehan
Institute of Telematics, University of Lübeck, Lübeck, Germany
e-mail: rehan@itm.uni-luebeck.de

O. Chughtai
Electrical Electronics Engineering Department, Universiti Teknologi PETRONAS,
Seri Iskandar, Malaysia

© Springer Science+Business Media Singapore 2015                                     75
A. Laouiti et al. (eds.), *Vehicular Ad-hoc Networks for Smart Cities*, Advances
in Intelligent Systems and Computing 306, DOI 10.1007/978-981-287-158-9_7

(R-OB-VAN) and ZoomOut HELLO, and implemented them in MATLAB. The results show that ZoomOut HELLO generates least number of RREQ rebroadcast messages and has minimum network delay due to RREQ messages.

**Keywords** VANET · AODV · PGB · R-OB-VAN · ZoomOut HELLO · QoS routing

## 1 Introduction

Vehicular ad-hoc network (VANET) is a high speed mobile ad-hoc network (MANET) which operates in cities or on highways. In order to be part of a VANET, vehicles must be equipped with wireless transceivers and protocol stack to allow them to act as network nodes. IEEE 802.11p or dedicated short range communication (DSRC) is a standard wireless communication channel for use among VANET nodes. VANET nodes normally operate in two modes: vehicle to vehicle (V2V) and vehicle to infrastructure (V2I).

ZoomOut HELLO technique relieves routing protocol in terms of providing it a QoS open network that inherently has least end-to-end delay and data traffic disruption due to controlled control traffic.

VANET has special characteristics like dynamic topology due to high speed, frequent obstructions like buildings, trees, mountains, variable traffic density and limited mobility variation subject to roads. Therefore, MANET routing protocols are not suitable for VANET. Several unicast routing protocols have been developed specifically for VANETs. Position-based routing (PBR) or geographic routing (GR) is more suited for VANET as it uses the physical location of the vehicles on road and is therefore more scalable than MANET routing protocols [1]. Conventional MANET protocols are either address-based or topology-based, therefore. They cannot guarantee unique vehicle addresses for the entire VANET fleet and therefore prove to be less suitable for VANET [1]. Authors in [2–4] state that GR protocols are more suitable to highway VANET than topology based routing protocols. In PBR [1], a vehicle obtains its neighbour information using periodic 1-hop HELLO messages while that of destination is obtained through the use of location service. Most GR protocols focus on paths which exist in maps but do not take into consideration the fact that whether source to destination path will have vehicles on it. The examples of GR protocols which possess above features are [5–11].

## 2 Motivation

As stated above, geographic routing protocols use 1-hop HELLO broadcast as the "I am alive" message. It is used to advertise the presence of a vehicle between 100 ms and 1 s depending upon the technique in use. Even if data is not being transferred, HELLO messages are still sent periodically to update neighbourhood. When a routing protocol is in action, there are normally two tables: the neighbour table and the routing-cum-forwarding table. If beaconing, route discovery and data exchange are on the same channel (classical method), then data exchange is disrupted due to beacons and route discovery messages. If beaconing is on different channel and route discovery and data exchange are together on the other channel, then the data exchange is disrupted due to route discovery and route maintenance messages. We can minimize the delay caused during data exchange by reducing the number of destination discovery rebroadcasts. This will ultimately increase network throughput.

Our view is slightly different. We argue that, if neighbour discovery (local view of network) is performed carefully in an intelligent way and the steps performed in neighbour discovery are not repeated by the routing protocol, then route discovery can be performed in the quickest possible way. This can easily be achieved, if vehicles while exchanging beacons select some vehicles as relatives out of the 1-hop neighbours. The criteria of selecting relatives can be relative speeds, relative distances, direction and inter-alia. Now, when a routing message will be received, only that vehicle towards the destination will re-broadcast which will be a relative to the predecessor vehicle, where as other vehicles will discard the packet. In this way, ZoomOut model will always attempt to maintain highly connected network among 1-hop neighbours and this connected chain will start from one end of VANET fleet and terminate at the network disconnection point. This VANET model will ultimately prove to be more reliable and faster to disseminate information and to discover destination. The routing broadcasts will be reduced significantly in a more reliable way and with less collision.

## 3 Proposed ZoomOut HELLO Technique

In principal, ZoomOut HELLO (ZOH) is a 1-hop beaconing technique. It exploits wireless broadcast advantage in a distributive manner so that routing protocols running on top of it can get a guaranteed reliable network and feel a QoS open network. ZOH is designed to possess implicit features of high connectivity (also called reliability), least channel access delay during destination discovery, quick destination discovery and fast information dissemination.

### 3.1 Finding Relatives of a ZoomOut Vehicle

Front relative (FR) and behind relative (BR) are the two important features of ZOH as shown in Fig. 1. Typed HELLO messages are exchanged between 1-hop neighbours due to which FR and BR vehicles are selected by each zoomOut vehicle (Z). Information like vehicle ID, speed, position, direction, FR_ID and BR_ID are part of a ZOH message. Relative speed and the inter vehicle distance (IVD) between vehicle Z and its 1-hop neighbours are the key parameters while selecting FR and BR relatives. To select FR and BR relatives from the preferred front and behind regions of vehicle Z, we follow a rule and its two cases as stated below.

**Fig. 1** *Outer dotted circle* illustrates 1-hop neighbourhood of ZoomOut vehicle Z

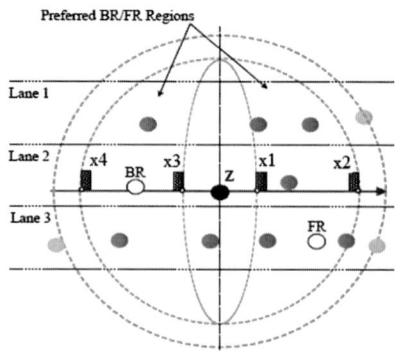

*Rule*: *If the connectivity of vehicle Z with a 1-hop neighbour B is longest in such a way that the hop count towards the destination is least while distance travelled is largest, then such 1-hop neighbour should become front or behind relative.*

**Case 1** When speed of vehicle Z is greater than the speed of its 1-hop neighbour, then a vehicle covering maximum distance becomes FR or BR and following relation is used.

$$D_b = \max_{n=1}^{m} \left[ S_z \times w \times \left( \frac{P_z - P_b}{S_z - S_b} \right) \right] \tag{1}$$

where $m$ are 1-hop neighbours having lower speed than vehicle Z; $w$ is weight in the range (0.1, 1); $P_z$, $S_z$, $P_b$, $S_b$ are the positions and speeds of vehicle Z and any of its 1-hop neighbour vehicle B respectively. Value of $w$ gives the time $t$ during which a portion of IVD is travelled.

**Case 2** When speed of vehicle Z is smaller than the speed of its 1-hop neighbour, then the vehicle covering maximum distance becomes BR or FR and following relation is used.

$$D'_b = \max_{n=1}^{m} \left[ S_z \times \left( \frac{r_z - (P_z - P_b)}{S_z - S_b} \right) \right] \tag{2}$$

where $r_z$ is the broadcast range of vehicle Z. Equation (2) gives time after which FR/BR relative will go out of range. $r_z = R - z$, where $R$ is broadcast range and $z$ represent a distance in metres away from broadcast boundary. A vehicle in $z$ region can disappear in a short time leaving unstable or disconnected VANET fleet.

If we zoomOut the broadcast regions and the whole VANET fleet, it follows that measures of relative speed and IVD give longer inter vehicle connectivity (IVC) within broadcast range (local scope) and a long lasting connected chain of relatives at highway (global scope).

## 3.2 Maintaining Relations with Relatives

Each ZOH vehicle maintains soft state with its relatives using typed HELLO messages as stated above. In Fig. 2, vehicle A and vehicle D exchange typed HELLO messages, T1 and T3, and maintain forward relation (FR), reverse relation (RR),

**Fig. 2** Sequence of FR-FRR-RR and RR-RBR-BR on two ZoomOut vehicles

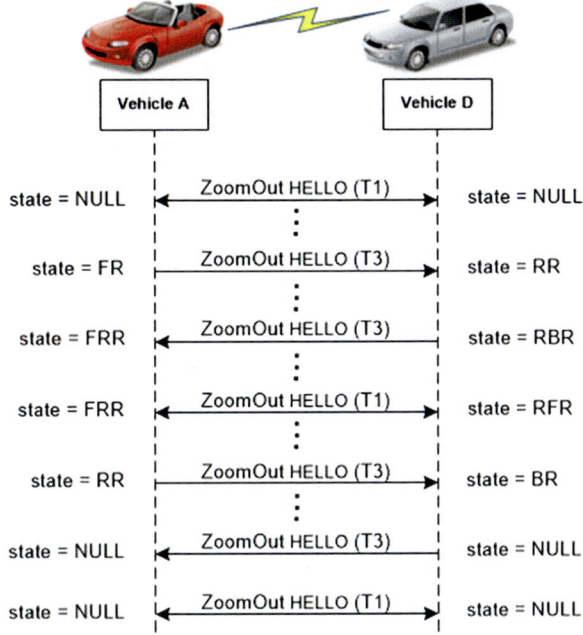

forward reverse relation (FRR) and reverse behind relation (RBR) and behind relation (BR). In Fig. 3 we present findRelatives() algorithm whose complexity is $O(n)$.

# 4 Analytical Model

Below we present an analytical model of traditional HELLO Ad-hoc on-demand distance vector (AODV) [12], PGB [13], R-OB-VAN [14] and ZoomOut techniques against two parameters: network delay and routing overhead.

In PGB [13], for every RREQ, each receiving vehicle has to classify itself with respect to received signal strength of sender. If the receiving node lies in the preferred group (PG), it can rebroadcast. Before, rebroadcast, the node in PG group has to wait for a hold off time interval to confirm that another belonging to PG has not broadcasted the RREQ packet. A node in the PG after receiving two RREQs during the hold off time will drop the RREQ. The concept of hold off timer adds further delay in the RREQs. The main issue here is that when the hold off time expires and the node decides to retransmit, the channel may be busy. The node has to wait for channel access. During the time, this node is waiting for random channel access; two other nodes rebroadcast RREQ. Now this node is unaware of it, so it will rebroadcast RREQ. Second, each PG that rebroadcasts adds additional information in the RREQ field of the routing protocol, which means a routing protocol on top of PGB, has to reserve extra fields so that it can function correctly.

```
void findRelatives() {
  For(; nbVehicle;nbVehicle=nbVehicle->nb_link.le_next){
    frontRelative = 0; behindRelative = 0;
    S_b = nbVehicle->speedOfNeighbour;  P_b = nbVehicle->X;
    if(S_z >= S_b) {
      relative_speed = S_z - S_b;
      if((relative_speed==0)||((relative_speed>0) && (rela-
      tive_speed<1))) relative_speed = 1;
      // Front Region of Vehicle Z
      if((Pb >= ((Px1-Pz)/2)) && (Pb <= Px1)) {
        tmpTime=(Pb-(Px1-((Px1-Pz)/2)))/relative_speed;
        if((NFR_slowFR_time==-1)||(NFR_slowFR_time < tmp-
        Time))        NFR_slowFR_time = tmpTime;
        NFR_slowFR = nbVehicle;
      } else if ((Pb > Px1) && (Pb < mid_Px1_Px2)) {
        tmpTime = (Pb - Px1)/relative_speed;
        if ((CFR1_slowFR_time == -1) || (CFR1_slowFR_time <
        tmpTime))  CFR1_slowFR_time = tmpTime;
        CFR1_slowFR = nbVehicle;
      } else if ((Pb >= mid_Px1_Px2) && (Pb < Px2)) {
        tmpTime = (Pb - mid_Px1_Px2)/relative_speed;
        if ((CFR2_slowFR_time == -1) || (CFR2_slowFR_time <
        tmpTime))  CFR2_slowFR_time = tmpTime;
        CFR2_slowFR = nbVehicle;
        } else if((Pb>=Px2) && (Pb<=((newR_front-Px2)/2))){
          tmpTime = (Pb - mid_Px1_Px2)/relative_speed;
          if ((FFR_slowFR_time == -1) || (FFR_slowFR_time <
          tmpTime)) FFR_slowFR_time = tmpTime;
          FFR_slowFR = nbVehicle;
      }}
  }
}
```

**Fig. 3** Find relatives algorithm finds FR and BR relatives among 1-hop neighbours

In R-OB-VAN [14], among 1-hop neighbours an acknowledgement scheme called active signalling phase takes place. After receiving a broadcast packet, the rebroadcast is done by a vehicle that is selected by the predecessor vehicle. All vehicles enter into random listening and produce an acknowledgement, which is basically the sequence of send and receive intervals. A send is represented by 1 and receive is represented by 0. The best progressive vehicle is selected and it then rebroadcasts. If a vehicle listens ACK from another vehicle, it infers that another better vehicle is available to rebroadcast, so it does not send its signal. The active signalling phase adds a delay of $t$ which is the sum of inter frame spacing (IFS = SIFS = 32 μs) and ACK (=SIFS + 1 Slot + transmission time (<4 μs) = 32 + 16 + 4= ~52 μs). Hence, per vehicle delay is ~84 μs.

## 4.1 Delay Analysis

We assume that there are total M nodes in the mobile ad-hoc network, which is VANET in this scenario. For a node, $i$, there are $m$ neighbour such that $m$ belongs to $\{1, 2, 3, \ldots, M\}$. First we calculate the delay and then broadcast routing overhead. The channel access delay can be expressed as:

$$T = t_{\text{DIFS}} + t_{\text{SIFS}} = 64 + 32 = 96 \ \mu s \tag{3}$$

**Network level delays due to traditional HELLO.** For one broadcast region, the delay to access the channel second time is sum of $m$ random channel access times $T$:

$$d_{\text{node}} = \sum_{j=1}^{m} T_j \tag{4}$$

For $N$ broadcast regions, delay from source to destination is:

$$N = (P_D - P_S)/2r \tag{5}$$

On highway, all vehicles will re-broadcast at least once due to small road width so:

$$d_{\text{traditional}} = \sum_{i=1}^{N} \sum_{j=1}^{m} T_{ij} \tag{6}$$

**Network level delays due to PGB.** A PGB vehicle adds an additional delay of random wait $t1$ before rebroadcasting as explained above. Delay of a PGB vehicle is $(P = p + t1)$, where $2 \leq p \leq 4$, so we replace $m$ in Eq. (4) with $P$ and re-write Eq. (6) as shown below:

$$d_{\text{PGB}} = \sum_{i=1}^{N} \sum_{j=1}^{P} T_{ij} \tag{7}$$

**Network level delays due to R-OB-VAN.** An R-OB-VAN vehicle adds an additional delay of random wait $t$ before rebroadcasting as explained above. Delay of an R-OB-VAN vehicle is $(O = m + t2)$, where $t2 = n * 84 \ \mu s)$, Here $n$ is the number of vehicles which will be involved in the active signalling. So we replace $m$ in Eq. (4) with $O$ and re-write Eq. (6) as shown below:

$$d_{\text{R-OB-VAN}} = \sum_{i=1}^{N} \sum_{j=1}^{O} T_{ij} \tag{8}$$

**Network level delays due to ZoomOut HELLO.** Since vehicle Z has two relatives $\{FR, BR\}$ so we replace $m$ in Eq. (4) with $z$ and re-write Eq. (6) as:

$$d_{\text{ZOH}} = \sum_{i=1}^{N} \sum_{j=1}^{z} T_{ij} \tag{9}$$

## 4.2 Route Request Rebroadcast Analysis

The RREQ rebroadcast is actually the same calculation but instead of time we count the number of re-broadcasts. We represent Eq. (9) for a general purpose RREQ re-broadcast calculation and replace the upper bound $q$ with $m$, $p$, $n$ or $z$ respectively for AODV (traditional HELLO), PGB, R-OB-VAN and ZoomOut HELLO.

$$\text{RREQ}_{\text{Total}} = \sum_{i=1}^{N} \sum_{j=1}^{q} B_{ij} \qquad (10)$$

# 5  Results

For MATLAB, we considered a VANET fleet on an area of 10 km having two-lanes in one direction. Since the range of IEEE 802.11p is ~1 km, so the broadcast range of every node is 1 km. The IVDs were set to 2, 5, 10, 20, 100 and 400 m for highly dense, dense, lightly dense, lightly sparse, sparse and highly sparse VANET types respectively. We performed ten experiments and then took an average value of all experiments for each VANET type.

Figure 4 shows that if a single source initiates route requests, then network delay on the is highest for PGB due to the value of random wait value for each group. The wait timer increases if vehicles in the OUT group find no vehicles in the PG group. Although number of nodes in AODV produce delay but since R-OB-VAN also incorporates the mechanism of active signalling and wait,

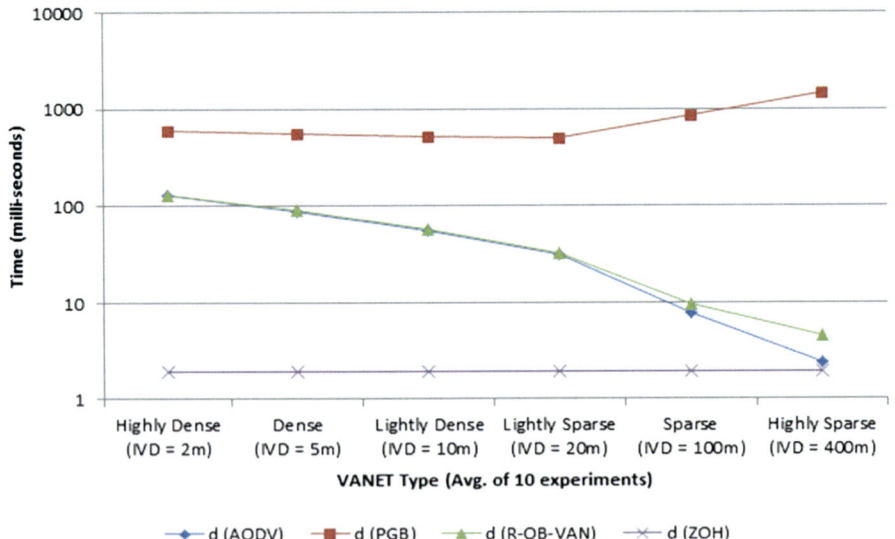

**Fig. 4** Comparison of average network delay due to RREQ re-broadcasts between ZoomOut and non-ZoomOut (AODV, PGB, R-OB-VAN) techniques

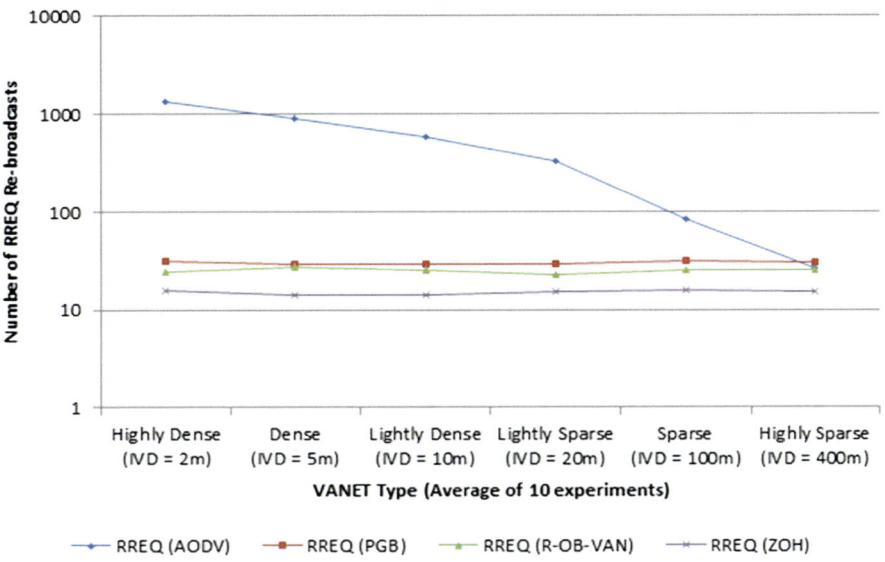

**Fig. 5** Comparison of the number of average RREQ re-broadcasts between ZoomOut and non-ZoomOut (AODV, PGB, R-OB-VAN) techniques

therefore, the two techniques fall in the same category except in dense VANET where AODV performs slightly better than R-OB-VAN. Since AODV does not have any timers so as soon as the number of vehicles decreases, the delay also increases in case of AODV but in case of R-OB-VAN it has to perform Acknowledgements to select the progressive vehicle, out of few vehicles, to disseminate information. ZoomOut only trusts in neighbours and therefore, it is almost negligibly affected even if VANET is highly dense or highly sparse.

Figure 5 shows that if a single source initiates a RREQ, the RREQs produced by AODV are highest as every node rebroadcasts at least once. The RREQ rebroadcasted by PGB are higher than the RREQ produced by R-OB-VAN. However, ZOH produces lest number of RREQ as only the FR or BR vehicle will re-broadcast based on the direction of packet.

## 6 Conclusion

In this work, we have shown that using ZoomOut HELLO beaconing we can further reduce routing overhead and network delay as compared to AODV, PGB and R-OB-VAN techniques. ZoomOut is a new 1-hop broadcast suppression technique designed to function below geographic routing protocols to assist them. The results show that if PBR protocols use ZOH, they will get least RREQ rebroadcasts and network delay. As a result, network throughput will also increase. In future, we will show how ZoomOut HELLO will achieve high degree of connectivity with its front and behind relatives.

# References

1. Zeadally, S., et al.: Vehicular ad hoc networks (VANETS): status, results, and challenges. Telecommun. Syst. **50**(4), 217–241 (2012)
2. Füßler, H., et al.: Mobicom poster: location-based routing for vehicular ad hoc networks. ACM SIGMOBILE Mobile Comput. Commun. Rev. **7**(1), 47–49 (2003)
3. Liu, G., et al.: A routing strategy for metropolis vehicular communications. In: Kahng, H.-K., Goto, S. (eds.) Information Networking. Networking Technologies for Broadband and Mobile Networks, pp. 134–143. Springer, Heidelberg (2004)
4. Menouar, H., Lenardi, M., Filali, F.: Movement prediction-based routing (MOPR) concept for position-based routing in vehicular networks. In: IEEE 66th Vehicular Technology Conference, 2007 VTC-2007 Fall. IEEE, New York (2007)
5. Jing, Z, Guohong, C.: VADD: vehicle-assisted data delivery in vehicular ad hoc networks. In: Proceedings of 25th IEEE International Conference on Computer Communications (INFOCOM 2006)
6. Sun, W., et al.: GVGrid: a QoS routing protocol for vehicular ad hoc networks. In: 14th IEEE International Workshop on Quality of Service, 2006 (IWQoS 2006)
7. Leontiadis, I., Mascolo, C.: GeOpps: geographical opportunistic routing for vehicular networks. In: IEEE International Symposium on a World of Wireless, Mobile and Multimedia Networks, 2007 (WoWMoM 2007)
8. Naumov, V, Gross, T.R.: Connectivity-aware routing (CAR) in vehicular ad hoc networks. In: INFOCOM 2007. 26th IEEE International Conference on Computer Communications. IEEE, New York (2007)
9. Hui, C., et al.: GrLS: group-based location service in mobile ad hoc networks. IEEE Trans. Veh. Technol. **57**(6), 3693–3707 (2008)
10. Yong, D., Li, X.: SADV: static-node-assisted adaptive data dissemination in vehicular networks. IEEE Trans. Veh. Technol. **59**(5), 2445–2455 (2010)
11. Chen, Y.-S., Lin, Y.-W., Pan, C.-Y.: DIR: diagonal-intersection-based routing protocol for vehicular ad hoc networks. Telecommun. Syst. **46**(4), 299–316 (2011)
12. Perkins, C.E. et al.: Ad hoc on-demand distance vector (AODV) routing, internet engineering task force. http://www.ietf.org/rfc/rfc3561.txt, 37 p. (2003)
13. Naumov, V., Baumann, R., Gross, T.: An evaluation of inter-vehicle ad hoc networks based on realistic vehicular traces. In: Proceedings of the 7th ACM International Symposium on Mobile Ad hoc Networking and Computing, pp. 108–119. ACM, Florence, Italy (2006)
14. Laouiti, A., Muhlethaler, P., Toor, Y. Reliable opportunistic broadcast in VANETs (R-OB-VAN). In: 9th International Conference on Intelligent Transport Systems Telecommunications (ITST) (2009)

# Information Delivery Improvement for Safety Applications in VANET by Minimizing Rayleigh and Rician Fading Effect

Muhammad Sajjad Akbar, Amir Qayyum and Kishwer Abdul Khaliq

**Abstract** The fading environment plays a critical role for vehicular communication. High fading environments lead to high data losses, which seriously affect the communication of safety applications in vehicular ad-hoc networkings (VANETs). Rician and Rayleigh fading models are considered to be realistic fading models for vehicular communication. This paper contributes in two aspects specifically for safety applications (a) Evaluation of Rician and Rayleigh fading for vehicular environment (b) It proposes a suitable combination of data rate and transmission power (Tx) that will minimize fading effects caused by Rician and Rayleigh fading. NCTUns is used to simulate vehicles with a safety application protocol called wave short message protocol (WSMP). By using appropriate combination of these parameters, significant improvements have been observed in throughput.

**Keywords** VANET · LoS · Rayleigh · Rician · Fading

## 1 Introduction

Vehicular ad-hoc networking (VANET) is a wireless network domain that incorporates vehicular communication in a limited wireless range. VANET is emerged as a subset of mobile ad-hoc networks (MANET) [2]. VANET is different from MANET in terms of predictable movements and high speed mobility. Current IEEE wireless LAN standard 802.11p is generally recommended for vehicular communication and performs

M.S. Akbar (✉) · A. Qayyum · K.A. Khaliq
CoReNeT, Mohammad Ali Jinnah University, Islamabad, Pakistan
e-mail: sajjad@corenet.org.pk
URL: http://www.corenet.org.pk

A. Qayyum
e-mail: aqayyum@ieee.org

K.A. Khaliq
e-mail: kishwer.a.k@ieee.org

© Springer Science+Business Media Singapore 2015
A. Laouiti et al. (eds.), *Vehicular Ad-hoc Networks for Smart Cities*, Advances in Intelligent Systems and Computing 306, DOI 10.1007/978-981-287-158-9_8

well for emergency situations [1, 11]. IEEE 802.11p is different from traditional wireless LAN standards (IEEE 802.11 a/b/g) in terms of supporting high communication range (1,000 m) and better support for high mobility. IEEE 802.11p operates on 5.9 GHz band by supporting the idea of dedicated short range communication (DSRC) [12]. DSRC concept is approved by U.S federal communications commission (FCC) and spectrum contains six service channels (SCH) and one control channel (CCH). All the communication for safety application is done through control channel. For non-safety applications, service channels are used [10]. VANET primarily deals with safety applications; however, it also focuses on traffic management and non-safety applications. Vehicle to vehicle (V2V) [3] communication becomes important for the accidents occurring due to unknown road conditions. Using V2V communication, any vehicle can send emergency/warning messages to other vehicles without any delay.

For optimum use of channel resources and to provide desired quality of service (QoS) [15] support for upper layer applications, study of channel modeling is very important. Significant research is available to discuss different propagation models under MANET scenarios [5]. However, the study of propagation models, especially for VANET safety application is still a gray area of research.

This paper contributes to minimize the effect of high fading environment for VANET. Data rate and transmission power (Tx) are two important parameters in data transmission. Different values of these parameters highly affect the efficiency of the transmission as well as upper layer application. This research will identify the suitable values of these parameters to minimize the effect of high fading environment.

In this paper, we have evaluated Rician and Rayleigh fading models for VANET scenarios. To intelligently cater the fading effects, we have also proposed the different cross layer parameters, such as Data rate and transmission power. We have proved that significant improvements in QoS support can be achieved in throughput values by using appropriate combination of different propose parameters. Rest of the paper is organized as follows. Section 2 discusses the propagation models used for the VANET. Section 3 defines the problem and related work. Section 4 presents the simulation analysis, followed by conclusion.

## 2 Propagation Models for the VANET

With the start of network research, initially free space propagation was used by most of the research community due to its simplicity. Free space propagation is a very simple model as it just considers line of sight (LoS) as the main parameter. However, no LoS directly leads to communication failure. So, it just fantasizes the ideal channel conditions, which are not a very true case in a mobile network. The phenomenon magnifies specifically in VANET where vehicles move on the road and have different obstacle on the way. Roadside buildings and on road obstacles create a lot of reflection making this propagation model not suitable for VANET.

Two ray ground radio (TRG) propagation model is also very popular and the most used propagation model for the wireless transmission. It is also the default propagation model for network simulator NS-2. Making it more realistic than free space fading model, TRG model considers two propagation parameters, i.e., LoS and the single reflection from the ground. TRG propagation model proved to be a good propagation model for the MANETs as the nodes appear relatively close to each other. Moreover, due to peculiar design, practical communication range of MANET nodes is much lower than the offered by MAC protocol. However, the presence of multiple reflections offers serious limitations for the use of TRG in VANETs. In VANETs, specifically in urban areas, vehicles freely move on the network of roads. There can be different types of obstacles (trees and buildings, etc.), which can disturb the signal strength against signal noise and interference. The TRG propagation model does not consider these factors as it only considers the LoS and ground-reflected signals. However, the TRG model can be proved as partially realistic for highway scenarios, where surroundings are clear from signal reflecting obstacles. However, precise fading conditions cannot be evaluated without considering multipath reflections.

Rician and Rayleigh are the two signal propagation models that consider LoS and multiple reflected signals [6]. In the presence of multiple multipath reflections, Rician fading model considers LoS as a dominant factor for the fading calculations. Lack of signal reflecting obstacles and the presence of straight road in open environment are the typical conditions for rural area or highway VANET communication. This phenomenon makes the Rician fading model as a simple but highly realistic model for the urban VANET scenarios.

On the other hand, Rayleigh fading model gives the due consideration to the LoS signals component as well as multipath reflections. The presence of big number of signal reflecting obstacles and limited presence of LoS component due to road turns and unbalanced heights are the base line conditions for urban area communication. These limitations make the relatively simpler Rayleigh fading model reasonably realistic for practical calculations.

## 3 Problem Statement

VANET communication among vehicles in highly fading environment (i.e. urban areas where roads are surrounded by the buildings) is a challenging task. The fading environment plays a critical role for data transmission and overall QoS support. High fading environment refers to low signal to interference and noise ratio (SINR). Low SINR leads toward high data loss and makes the communication unstable. In VANET, vehicles usually broadcast emergency messages in a limited area. High fading environment can disturb this communication and the results can be highly harmful, especially for time sensitive safety applications.

In VANET, Rician and Rayleigh fading models are more realistic than the free space or TRG propagation models. Different studies have used different propagation models in VANET. In the following, we have mentioned some of those studies.

The authors [9] have identified the limitations of TRG propagation model in highly fading environment. Comparing different parameters, authors proposed to use Nakagami propagation model for highly fading environment. They used two different routing protocols, i.e., OLSR [8] and AODV [4, 14] and analyzed the performance under Nakagami propagation model. OLSR performed better than AODV in terms of packet delivery ratio. To make the simulation simple, the authors ignored the speed of vehicles in the mobility model and used a constant value of the speed for the vehicles. Many authors [13] have studied the different propagation models with their suitability in VANET. These propagation models include both realistic models and the deterministic models. They proposed to use realistic propagation model in highly fading environment, specifically Rician, Rayleigh, and Nakagami. The authors only gave their analysis without any simulation results. So, it is not easy to make a concrete conclusion. Further, they have not mentioned any specific scenario to support their outcomes.

Another study [5] has discussed the urban and highway environment for DSRC with Nakagami propagation model. However, authors have not mentioned any mobility model for the simulation scenario. Lack of mobility model makes the conclusion quite complicated.

In addition to these studies, there are different studies available for Rician and Rayleigh fading models for VANET [10]. However, results computed by these studies at basic level make them less useful for realistic environments. Moreover, in many such studies, authors have not included the enough information for the topology.

## 4 Simulation and Analysis

In the next section, we discuss our purposed classifications of traffic under urban as well as highway scenarios. For the test environment, we have defined a VANET topology with both urban and highway scenarios. Different test data types were generated to observe the behavior of fading environments. This paper shows multiple simulations including broadcast scenarios for Wave Short Message Protocol (WSMP) with variation in some physical layer parameters like data rate and transmission power. All simulations were tested separately for Rayleigh and Rician fading models depicting urban and highway scenarios.

As a network simulation tool, we selected NCTUns 5.0 an open source network simulator. Few studies [7] have proved the efficiency of NCTUns over other open source simulators like *NS*-2 and *NS*-3. NCTUns provides support for WSMP with the application name WSMP_Forwarding. It also offers use of MAC protocol IEEE 802.11p.

Figure 1 shows the network topology for our simulation. In all the simulation scenarios, we considered a static node connected through a wired link to a road side unit (RSU). We generated different data traffics for the RSU over UDP depicting emergency management messages. Our simulation involves 20 vehicles as on board unit (OBU) moving with different speeds. The RSU broadcast the information received from static node by using WSMP_Forwarding function of NCTUns. We set the initial packet size of 1,400 bytes. Duration of the simulation was set to 35 s.

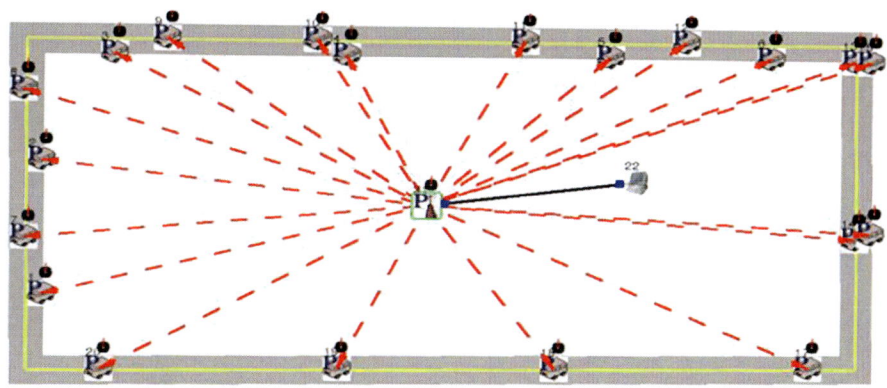

**Fig. 1** Simulation scenario

We assumed that WSMP application is running on all the vehicles. As an initial mobility model, we positioned each vehicle in range of RSU. We computed the data throughput at each vehicle while moving it away from RSU. Subsequently, we computed overall throughput of the topology for all scenarios. We computed the MAC layer throughput by using generate throughput log file option in NCTUns. We generated the relevant graphs through graphs generation option in NCTUns.

Table 1 provides the overall three possible values for data rate and transmission power. The detail analysis for the best results regarding throughput is discussed in subsequent paragraphs. We can observe that, whenever Tx is increased, we did not get the stable throughput results. Therefore, high Tx should be avoided for Rician fading environments.

**Table 1** Throughput calculation for Raleigh and Rician

| Fading models | Data rate (Mbps) | Tx (dBm) | Number of vehicles (V) | Throughput (Kbps) |
|---|---|---|---|---|
| Rayleigh | 6 | 15 | 20 | Fluctuate: 24–28 |
| Rician | 6 | 15 | 20 | Constant: 29 |
| Rayleigh | 6 | 28 | 20 | Fluctuate: 30–62 |
| Rician | 6 | 28 | 20 | Stable after few seconds and provides 33 |
| Rayleigh | 12 | 15 | 20 | Fluctuate: 19–27 |
| Rician | 12 | 15 | 20 | Stable after few seconds and provides 52 |
| Rayleigh | 12 | 28 | 20 | Fluctuate: 20–60 |
| Rician | 12 | 28 | 20 | Fluctuate: 30–60 |
| Rayleigh | 27 | 15 | 20 | Fluctuate: 15–30 |
| Rician | 27 | 15 | 20 | Stable after few seconds and provides 28 |
| Rayleigh | 27 | 28 | 20 | Fluctuate: 15–70 |
| Rician | 27 | 28 | 20 | Fluctuate: 20–60 |

**Fig. 2** Rician fading with
data rate 12 Mbps and
Tx 15 dBm

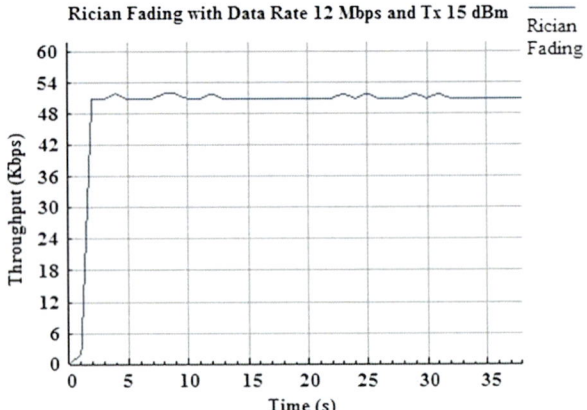

We initially studied Rician fading environment. We applied different combina-tions of data rate and transmission power (Tx) to check their effects on through-put. Figure 2 shows the combination of the data generation rate of 12 Mbps and Tx power of 15 dBm. We observed the maximum throughput gain up to 52 kbps. Whereas, with other combinations of data rate and transmission power the result-ing throughput remained much lower than 52 kbps. Hence, we recommend the use of 12 Mbps and Tx power of 15 dBm for Rician fading model to get realistic and better throughput.

Similar to Rician fading environment, we applied different combinations of data rate and transmission power (Tx) for Rayleigh fading environment. To check the effects on throughput, we observed that at data rate of 12 Mbps and even by increasing Tx to 28 dBm, throughput fluctuated between 7 and 62 kbps. The results are shown in Fig. 3. We can also observe that peak value in Rayleigh fading increased from Rician fading values. However, fluctuation in the response caused serious degradation over longer duration.

**Fig. 3** Rayleigh fading with
data rate 12 Mbps and
Tx 28 dBm

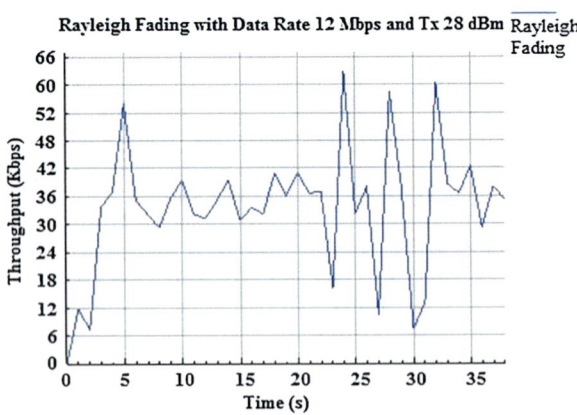

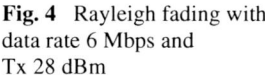

**Fig. 4** Rayleigh fading with data rate 6 Mbps and Tx 28 dBm

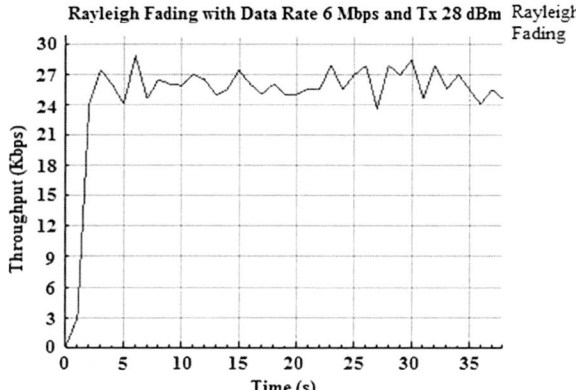

From Fig. 4, it is observed that contrary to throughput of 12 Mbps in Rician fading environment, the data rate of 6 Mbps under Tx value of 15 dBm, provided much stable result. However under such conditions, we only got maximum throughput of 28 kbps.

Hence, with this simple and realistic study, we can conclude that same data rates cannot be achieved for same network conditions under Rayleigh and Rician fading environment. In other words, we can say that a vehicle node may face a sudden degradation of half of throughput while entering in urban area from a highway. This degradation requires a dedicated planning while offering QoS sensitive applications. However, considering the higher node density under urban environments, a node may face even much higher degradation under urban scenarios.

The prior planning and availability of roadside infrastructure is of utmost importance for smooth QoS provisioning under combination of urban and highway scenarios. We can choose the best combination of data rate and transmission value in a fading environment and leads toward less data loss.

## 5 Conclusion

For VANET, Rician and Rayleigh are considered as realistic fading environment. High fading environment causes high data loss. As a result, a vehicle achieves much throughput value and fluctuating QoS support. To minimize the fading effects, efficient selection of data rate and transmission power (Tx) are considered as important transmission parameters. Appropriate value combination of these parameters minimize the fading effect significantly. In the simulation, we have considered different combination values of the parameters for Rician and Rayleigh fading environments. We tried to find out the best combination of these parameters. For Rician fading, the combination of the data rate 12 Mbps and Tx 15 dBm gave the high and stable throughput values. However for the Rayleigh fading, the combination of the data rate 6 Mbps and Tx 15 dbm achieved the stable through-put. Hence, these combinations are considered as best combination to minimize the fading effect in high fading environments. Hence, selection of appropriate values of these parameters is very important in VANET. Our research contributes in minimizing the effect of data loss in high fading environment and guide toward high throughput.

# References

1. Amadeo, M., Campolo, C., Molinaro, A.: Enhancing IEEE 802.11 p/WAVE to provide info-tainment applications in VANETs. Ad Hoc Netw. **10**(2), 253–269 (2012)
2. Bansal, M., Rajput, R., Gupta, G.: Mobile ad hoc networking (MANET): routing protocol performance issues and evaluation considerations. Tech. rep., RFC 2501 (Informational) (1999)
3. Biswas, S., Tatchikou, R., Dion, F.: Vehicle-to-vehicle wireless communication protocols for enhancing highway traffic safety. IEEE Commun. Mag. **44**(1), 74–82 (2006)
4. Chakeres, I.D., Belding-Royer, E.M.: AODV routing protocol implementation design. In: Proceedings of 24th International Conference on Distributed Computing Systems Workshops 2004, pp. 698–703 (2004)
5. Dahiya, A., Chauhan, R.: A comparative study of MANET and VANET environment. J. Comput. **2**(7), 87–92 (2010)
6. Goldsmith, A.: Wireless Communications. Cambridge University Press, Cambridge (2005)
7. Hassan, A.: VANET simulation. Ph.D. thesis, Halmstad University (2009)
8. Jacquet, P., Muhlethaler, P., Clausen, T., Laouiti, A., Qayyum, A., Viennot, L.: Optimized link state routing protocol for ad hoc networks. In: Proceedings of IEEE International Multi Topic Conference, IEEE INMIC 2001. Technology for the 21st Century, pp. 62–68 (2001)
9. Khan, I., Qayyum, A.: Performance evaluation of AODV and OLSR in highly fading vehicular ad hoc network environments. In: IEEE 13th International Multitopic Conference, INMIC 2009, pp. 1–5 (2009)
10. Liu, B., Khorashadi, B., Du, H., Ghosal, D., Chuah, C.N., Zhang, M.: VGSim: an integrated networking and microscopic vehicular mobility simulation platform. IEEE Commun. Mag. **47**(5), 134–141 (2009)
11. Mohammad, S.A., Rasheed, A., Qayyum, A.: VANET architectures and protocol stacks: a survey. In: Communication Technologies for Vehicles, pp. 95–105. Springer, Berlin (2011)
12. Morgan, Y.L.: Notes on DSRC and WAVE standards suite: its architecture, design, and characteristics. IEEE Commun. Surv. Tutorials **12**(4), 504–518 (2010)
13. Singh, P.K., Lego, K.: Comparative study of radio propagation and mobility models in vehicular ad hoc network. Int. J. Comput. Appl. **16**(8), 37–42 (2011)
14. Royer, E.M., Perkins, C.E.: An implementation study of the AODV routing protocol. In: Wireless Communications and Networking Conference on IEEE, WCNC 2000, vol. 3, pp. 1003–1008 (2000)
15. Xiao, H., Seah, W.K., Lo, A., Chua, K.C.: A flexible quality of service model for mobile ad hoc networks. In: Vehicular Technology Conference Proceedings IEEE 51st, 2000, VTC 2000-Spring Tokyo, vol. 1, pp. 445–449. (2000)

# Author Index

© Springer Science+Business Media Singapore 2015                    93
A. Laouiti et al. (eds.), *Vehicular Ad-hoc Networks for Smart Cities*, Advances
in Intelligent Systems and Computing 306, DOI 10.1007/978-981-287-158-9